SYSTEMATIC BOTANY MONOGRAPHS

VOLUME 24

Systematics of Oenothera Section Oenothera Subsection Raimannia and Subsection Nutantigemma (Onagraceae)

Werner Dietrich
Warren L. Wagner

THE AMERICAN SOCIETY OF PLANT TAXONOMISTS

8 August 1988

SYSTEMATIC BOTANY MONOGRAPHS
ISSN 0737-8211

ISBN 0–912861–24–X

Printed in the United States of America

Volume 24 of Systematic Botany Monographs is published with the support of National Science Foundation Grant No. BSR-8714636. The Foundation provides awards for research and education in the sciences. The awardee is wholly responsible for the conduct of such research and preparation of the results for the publication. The Foundation, therefore, does not assume responsibility for such findings or their interpretation. Any opinions, findings, conclusions, or recommendations expressed in this publication are those of the author and do not necessarily reflect the views of the National Science Foundation.

SYSTEMATICS OF OENOTHERA SECTION OENOTHERA SUBSECTION RAIMANNIA AND SUBSECTION NUTANTIGEMMA (ONAGRACEAE)

Werner Dietrich
Botanisches Institut der Universität Düsseldorf
Universitätsstr. 1
D–4000 Düsseldorf, Germany

Warren L. Wagner
Department of Botany
Bernice P. Bishop Museum
Honolulu, Hawai'i 96817

Abstract. This is the third paper in a series treating the five subsections of the large *Oenothera* sect. *Oenothera* (76 species), divided into its present form based on comparative morphological studies and on genome and plastome relationships inferred from an extensive crossing program. The taxonomy, details of breeding systems, cytology, and a discussion of the relationships pertinent to the classification of the species of subsect. *Raimannia* (11 species) and subsect. *Nutantigemma* (3 species) are presented. The five species of subsect. *Raimannia* having erect stems, compact densely flowered inflorescences, and lanceoloid capsules are placed in ser. *Candela. Oenothera heterophylla, O. cordata,* and *O. rhombipetala* have the plesiomorphic features of large flowers and self-incompatibility. These species, which occur from eastern Texas north throughout the Great Plains, form bivalents or small translocation rings and bivalents in meiosis. The other two species, *O. clelandii* and *O. curtissii,* appear to have been independently derived from *O. rhombipetala,* but are now geographically removed from their progenitor. Both are small-flowered, autogamous, permanent structural heterozygotes, forming a ring of 14 chromosomes in meiosis and having about 50% pollen fertility. *Oenothera clelandii* occurs north and east of *O. rhombipetala* in the central lowlands, from eastern Minnesota and eastern Iowa to northern Indiana and Michigan, and *O. curtissii* is restricted to scattered localities in the southeastern U.S.A. (Alabama, Florida, Georgia, and South Carolina). The remaining six species of subsect. *Raimannia* are placed in ser. *Raimannia,* characterized by relatively few-flowered inflorescences with lateral branches and cylindrical capsules. This group is centered in the southern grasslands of Texas, where four of the six species occur. *Oenothera grandis,* primarily of the Great Plains to southern Texas, is the only self-incompatible species of the series. *Oenothera falfurriae* and *O. mexicana,* endemics of southern Texas, are both small-flowered, autogamous bivalent formers. *Oenothera laciniata,* which occurs nearly throughout the eastern U.S.A. and is widely naturalized, is an autogamous, permanent structural heterozygote derived from *O. grandis.* The other two species of ser. *Raimannia* are restricted to coastal dune areas on the Gulf of Mexico and the Atlantic. *Oenothera drummondii* subsp. *drummondii,* a large-flowered, self-compatible, modally outcrossing subspecies, occurs from Campeche, Mexico, to North Carolina but is rare and apparently relictual from Louisiana to North Carolina. It is replaced by the derived permanent structural heterozygote *O. humifusa* from Louisiana northward to Pennsylvania. *Oenothera drummondii* subsp. *thalassaphila* is a rare self-compatible, modally outcrossing, bivalent former disjunct in Baja California Sur. Subsection *Nutantigemma* consists of three species that exhibit genetic differentiation; they form sterile hybrids with irregular meiosis in crossing experiments with members of subsect. *Raimannia. Oenothera pubescens,* a permanent structural heterozygote, has the widest natural distribution in the genus: from the southwestern U.S.A. to Andean South America and the Galapagos Islands. Also included in this subsection are *O. breedlovei,* endemic to Sierra Laguna, Baja California Sur, and *O. tamrae,* known only from the type collected in Nayarit, Mexico. *Oenothera breedlovei* is a large-flowered, self-compatible, presumably modally outcrossing, relict species; *O. tamrae,* with large flowers and pollen fertility of over 90%, is also presumed to be a bivalent former.

INTRODUCTION

This is the third paper in a series treating the five subsections of the large section *Oenothera,* which consists of 76 species. The section was circumscribed in its present form by Stubbe and Raven (1979), based on genome and plastome relationships inferred from an extensive crossing program in Düsseldorf and morphological studies. Subsection *Munzia* W. Dietrich, widespread in South America and the largest group consisting of 45 species, was treated in detail by Dietrich (1977). More recently (Dietrich et al. 1985) a revision of subsect. *Emersonia* (Munz) W. Dietrich, Raven & W. L. Wagner was completed. This subsection consists of four species which have scattered relictual distributions in central and northern Mexico and New Mexico, and it may represent the earliest lineage to have evolved in the section. The present paper presents the taxonomy and a discussion of the relationships of the species of subsect. *Raimannia* (Rose ex Britton & A. Brown) W. Dietrich (11 species) and subsect. *Nutantigemma* W. Dietrich & W. L. Wagner (3 species). This leaves only subsect. *Oenothera* (13 species) within sect. *Oenothera* without a modern treatment.

Past efforts (Séringe 1828; Fischer & Meyer 1835; Spach 1835; Rose 1905; Munz 1935, 1965) have not lead to a completely satisfactory taxonomy of much of *Oenothera,* especially sect. *Oenothera,* primarily because they were based almost exclusively on morphological observations alone. Recent experimental crossing analyses and cytological studies, coupled with comparative morphological studies, have given new insight into the relationships in *Oenothera* (Raven 1964, 1969; Towner 1977; Dietrich 1977; Dietrich et al. 1985; Straley 1977; Raven et al. 1979; Stubbe & Raven 1979; Wagner et al. 1985). Use of the methods of crossing analysis established for *Oenothera* primarily by Renner, Cleland, and Stubbe (see summary in Cleland 1972), involving chromosome pairing in meiosis, the fertility of hybrids, and the compatibility and level of genome/plastome interaction, has made it possible to study critically the relationships of species or groups of species.

This paper is based on the study of extensive collections of 14 species throughout their natural range, made possible largely by the initiative of Peter H. Raven. These strains were cultivated at the Botanical Institute of the University of Düsseldorf. During the past 25 years collections from a total of 150 localities of the 14 species constituting subsections *Raimannia* and *Nutantigemma* have been studied. These experimental garden studies coupled with studies of comparative morphology and breeding systems formed the basis of the taxonomy presented here. In 1978 Dietrich, at the herbarium of the Missouri Botanical Garden, examined approximately 6,000 specimens loaned from primarily North American herbaria. This work supplemented the experimental studies for comparative morphology and patterns of variation. The collection data were used in the preparation of the distribution maps. This work led to the reinstatement of *Oenothera cordata, O. curtissii,* and *O. mexicana,* which have been recognized by few researchers since their original description (e.g., Munz 1965), as well as to the discovery of four new species, *O. clelandii, O. falfurriae, O. breedlovei,* and *O. tamrae,* described elsewhere (Wagner

1983; Dietrich & Wagner 1987). Despite intensive searching by D. E. Breedlove, the last species could not be relocated where it was originally collected in Nayarit, Mexico, and thus could not be studied experimentally. On the other hand, the search for *O. tamrae* led to the discovery of a new locality for *O. pubescens,* the only widespread member of subsect. *Nutantigemma* and a permanent translocation heterozygote.

CYTOLOGY AND BREEDING SYSTEMS

Determination of chromosome number, meiotic configurations, and self-compatibility by repeated self-pollination was made on all of the strains that have been brought into cultivation, which represent 13 species (including four subspecies) from 150 localities. The meiotic behavior of a total of 227 individual plants was studied; subdivided they represent 39 plants from 27 localities of ser. *Candela,* 162 plants from 102 localities of ser. *Raimannia,* and 26 plants from 21 localities of subsect. *Nutantigemma.* All counts were diploid, n = 7, with no polyploidy or aneuploidy detected. Chromosome configurations, and breeding systems for subsections *Raimannia* and *Nutantigemma* are summarized in Table 1; the details for each taxon are given in the taxonomic treatment.

Within ser. *Candela,* the large-flowered species with stigmas elevated above the anthers at anthesis—*Oenothera heterophylla, O. cordata,* and *O. rhombipetala*—are bivalent-forming species, with 7_{II} or small rings of up to a few reciprocal translocations at meiotic metaphase I; their pollen is usually about 90–100% fertile. The occasional occurrence of plants of these species that have one to three reciprocal translocations, forming small rings of chromosomes in meiosis, represents a situation that is not uncommon in species of tribe Onagreae (Raven 1979). With the exception of the Alabama strains of *O. heterophylla* subsp. *orientalis,* these three species are self-incompatible obligate outcrossers (see also Cleland 1960). The small-flowered *O. clelandii* and *O. curtissii* are both modally autogamous, permanent structural heterozygotes; that is, they form a ring of 14 chromosomes at meiotic metaphase I and have about 50% fertile pollen.

These two permanent structural heterozygotic species do not appear to fit Renner's concept (1917) of the evolution of permanent heterozygosity as formulated for species of subsect. *Oenothera.* He proposed that a permanent heterozygote evolves from a hybridization event between two different species with unique phenotypic differences. In the case of both *O. clelandii* and *O. curtissii,* hybrids with structural homozygotic species showed only minor phenotypic differences. It thus appears that both complexes of these two permanent structural heterozygotic species arose from within a single species, *O. rhombipetala.* It has been suggested (Stubbe 1980; Dietrich 1977; Wasmund 1980, 1984) that gradual accumulation of translocations is the most likely explanation for the evolution of certain permanent structural heterozygotic species of subsect. *Oenothera* or subsect. *Munzia.* These are species homozygous for the complexes they carry, such as the AA permanent heterozygote *O. villosa* Thunberg or the BB permanent heterozygote *O. nutans* Atkinson & Bartlett (both of subsect. *Oenothera*). This model may explain the evolution of *O. clelandii* and *O. curtissii* from ancestral populations similar to *O. rhombipetala.* It is also possible that they evolved as the result of an interpopula-

Table 1. Summary of chromosome configurations and breeding systems in *Oenothera* sect. *Oenothera* subsect. *Raimannia* and subsect. *Nutantigemma.* The following symbols are used: II = bivalents; ⊙ = translocation ring; SI = self-incompatible; SC = self-compatible; MO = modally outcrossing; MA = modally autogamous; PTH = permanent translocation heterozygote.

	Meiotic chromosome configurations	Breeding systems
Subsect. *Raimannia*		
Series *Candela*		
O. heterophylla		
subsp. *heterophylla*	7_{II}; 5_{II} + ⊙4; 3_{II} + 2⊙4	SI
subsp. *orientalis*	7_{II}	SI (Arkansas) SC (Alabama), MO
O. cordata	7_{II}; 5_{II} + ⊙4	SI
O. rhombipetala	7_{II}; 5_{II} + ⊙4; 1_{II} + 3⊙4	SI
O. clelandii	⊙14	SC, MA, PTH
O. curtissii	⊙14	SC, MA, PTH
Series *Raimannia*		
O. grandis	7_{II}; 5_{II} + ⊙4	SI
O. falfurriae	7_{II}	SC, MA
O. mexicana	7_{II}	SC, MA
O. laciniata	⊙14	SC, MA, PTH
O. drummondii		
subsp. *drummondii*	7_{II}; 5_{II} + ⊙4; 4_{II} + ⊙6; 3_{II} + ⊙8	SC, MO
subsp. *thalassaphila*	7_{II}	SC, MO
O. humifusa	⊙14	SC, MA, PTH
Subsect. *Nutantigemma*		
O. breedlovei	7_{II}	SC, MO
O. tamrae	—	—, probably MO
O. pubescens	⊙14; 1_{II} + ⊙12	SC, MA, PTH

tional hybrid event in the manner suggested by Ellstrand and Levin (1980c) for *O. laciniata.* Current data do not allow us to suggest that one hypothesis is more likely than the other. Holsinger and Ellstrand (1984) compare basic hypotheses concerning evolution of permanent structural heterozygosity.

Within ser. *Raimannia, O. grandis, O. falfurriae, O. mexicana,* and both subspecies of *O. drummondii* are bivalent-forming taxa, whereas *O. laciniata* and *O. humifusa* are permanent structural heterozygotic species. *Oenothera grandis* is self-incompatible, but the other three bivalent-forming species are self-compatible. Also, the haploid genomes (complexes) of *O. laciniata* and *O. humifusa* differ only slightly from each other. Based on their analysis of allozyme data, Ellstrand and Levin (1980c) suggested that *O. laciniata* evolved as an interpopulational hybrid of *O. grandis*.

Of the three species of subsect. *Nutantigemma,* only *O. breedlovei,* and *O. pubescens* are presently in cultivation. The former species has large flowers and long styles with the stigma elevated above the anthers at anthesis. It is self-compatible but presumably modally outcrossing and is a structural homozygote. *Oenothera pubescens,* on the other hand, is modally autogamous and a permanent structural heterozygote; it is the most widespread species of the group. *Oenothera tamrae,* known only from a single collection made in Nayarit, Mexico (*Rose 2133*), has large flowers and the stigma is elevated above the anthers at anthesis, which suggest that it is also at least modally outcrossing, if not genetically self-incompatible. Large-flowered outcrossing *Oenothera* species nearly always form bivalents or small translocation rings as we would thus expect *O. tamrae* to do. That *O. tamrae* is probably not a permanent structural heterozygote is supported by the high percentage of fertile pollen (91%) in the type.

Ellstrand and Levin (1980a) studied the recombination system of three closely related, geographically and ecologically sympatric species of subsect. *Raimannia* in order to assess any differences between them. Their allozyme data showed that there were significant differences in the allozyme variation that correlated with the different recombination systems. The autogamous, bivalent-forming *O. mexicana* was monomorphic and displayed no detectable heterozygosity at any of the loci studied. In contrast, the outcrossing, self-incompatible, bivalent-forming *O. grandis,* as well as its closely related permanent structural heterozygote derivative *O. laciniata,* showed low levels of genetic diversity. Populations of *O. grandis* generally were close to Hardy-Weinberg expectations, whereas those of *O. laciniata* deviated strongly (Ellstrand & Levin 1980a, 1980b). Although the levels of variation species-wide were similar in *O. grandis* and *O. laciniata,* the latter had a significantly lower level of variation within populations; most of its variation was between populations, as would be expected of a permanent structural heterozygote (Ellstrand & Levin 1980a).

Ellstrand and Levin's analysis indicated that some intrapopulation variation exists. A survey of 60 populations and 18 enzyme loci (Ellstrand & Levin 1982) revealed a total of 108 genotypes in *O. laciniata.* Sixty-three genotypes were unique to only one population, and three were widespread. The number of genotypes recorded for a single population ranged from 1 to 16 with a mean of 6.5. *Oenothera laciniata* was considerably more diverse than some other animal and plant species with restricted recombination systems that have so far been studied (Ellstrand & Levin 1982). Hybridization with *O. grandis* was suggested by Ellstrand and Levin as a possible reason for the high levels of variation observed. Our studies suggest that *O. laciniata* hybridizes not only with *O. grandis,* but also with *O. mexicana, O. drummondii* subsp. *drummondii,* and *O. humifusa.* This may also contribute to the observed pattern of variation.

INFRAGENERIC CLASSIFICATION AND CROSSING RELATIONSHIPS

In an attempt to divide *Oenothera* into natural groups at the generic level, Rose (1905) established the genus *Raimannia* and designated *Oenothera laciniata* as the type. However, he did not provide a generic description, and the genus was not validly published until Britton and Brown (October 1913) provided the validating description. Also none of the names of the included species were validly published since the genus was not (ICBN Art. 43.1); some of the combinations were picked up by subsequent authors who thereby validated them in *Raimannia.* Rose also placed *O. coronopifolia* Torrey & A. Gray in *Raimannia;* however, this species, along with *O. albicaulis* Pursh, clearly differs in having white petals and subglobose pitted seeds. They are currently placed in the related sect. *Kleinia* Munz. Sprague and Riley (1921) enlarged *Raimannia* to include also what is now known as subsect. *Munzia* (Dietrich 1977) in South America.

Section *Oenothera* was circumscribed in its present form with five subsections by Stubbe and Raven (1979), who considered genome and plastome relationships and morphology. As is true of all other modern revisions in Onagraceae, all sectional realignments in *Oenothera* have been made using the narrow sectional concept that Lewis and Lewis (1955) established for the genus *Clarkia.* In this view, members of a section are closely related and share a large number of features. The sections now being established for *Oenothera* according to this concept largely conform to the subgenera of Munz (see his summary of 1965) with certain further subdivisions and a number of species realignments. Some of the sections in *Oenothera* can be grouped together into clear lineages, but the levels of distinction among them is not uniform (Wagner & Raven, unpubl.), and thus subgenera are not recognized at present.

The circumscription of sect. *Oenothera* has changed in a number of significant respects since the review of the genus in North America by Munz (1965). The species that we include here in subsect. *Raimannia* and subsect. *Nutantigemma* were included by Munz (1935, 1965) in his subg. *Raimannia.* His subgenus also included a number of other species that have since been transferred elsewhere. *Oenothera coronopifolia* and *O. albicaulis* were placed by Munz in a new section he described, sect. *Kleinia* of subg. *Raimannia.* The current view is that these two species are specialized members of this lineage, but are now considered to constitute a separate section; subg. *Raimannia* is no longer recognized. A recent genus-wide study of seed anatomy (Tobe et al. 1987) has shown that these two species closely resemble the species of subsect. *Raimannia* and subsect. *Nutantigemma* in this respect, but that they have several unique features and thus should be retained in the ditypic sect. *Kleinia.*

Raven (1970) removed *O. muelleri* and *O. macrosceles* from subg. *Raimannia* and placed them in Munz's subg. *Pachylophus,* which in Raven's view included eight species: *O. muelleri* Munz, *O. tubifera* Séringe, *O. xylocarpa* Coville, *O. primiveris* A. Gray, *O. caespitosa* Nuttall, *O. cavernae* Munz, *O. brandegeei* (Munz) Raven, and *O. macrosceles* A. Gray. Subsequent studies have suggested that these species, along with *O. organensis* Munz and *O. stubbei* W. Dietrich, Raven & W. L. Wagner, are indeed related, but that they can conveniently be subdivided into a number of monophyletic groupings, which we recognize as sections or subsections. Recent study of crossing relationships, morphology, and seed morphology (Stubbe & Raven 1979; Dietrich et al. 1985) showed that *O. macrosceles* is best grouped with *O. maysillesii* Munz, *O. stubbei,* and *O. organensis* in sect. *Oenothera* subsect. *Emersonia.* Related to subsect. *Emersonia* are *O. muelleri* and *O. tubifera,* which, based on similar types of studies, are now considered to constitute a ditypic section, sect. *Ravenia* W. L. Wagner (Wagner 1986). Section *Ravenia* is in turn the sister group to the remainder of the species listed above for subg. *Pachylophus,* which is now divided into sections *Eremia* W. L. Wagner, *Contortae* W. L. Wagner, and *Pachylophus* (Spach) Endlicher with the subgenus no longer recognized (Wagner et al. 1985; Wagner 1986).

Munz (1935) enlarged subg. *Raimannia* to include the usually yellow-flowered evening primroses widespread in South America, such as *O. stricta* Ledebour ex Link, the only South American species of *Oenothera* naturalized in North America, a decision based exclusively on his evaluation of their external morphology. Dietrich (1977) placed the South American species in subsect. *Munzia,* which is differentiated from the remainder of sect. *Oenothera* both by morphology and crossability (Dietrich 1977; Stubbe & Raven 1979). Therefore, subsect. *Raimannia* as recognized here is restricted to North America, except for *O. laciniata* and *O. drummondii* subsp. *drummondii,* which are naturalized in South America.

All artificial hybrids between species of subsections *Munzia* and *Raimannia* are pale green, which suggests that the plastids of one subsection do not function well with the genome of the other. These intersubsectional hybrids also have low fertility and have an irregular meiosis (Stubbe & Raven 1979). By contrast, hybrids among the species of subsect. *Raimannia* are completely fertile and have no disturbances or only minor ones during meiosis; however, some hybrid combinations exhibit plastid-genome incompatabilities and are thus pale or albino and also weak. Subsection *Raimannia,* including only the North American species, represents a natural group of closely related species that have changed relatively little genetically.

Oenothera pubescens recently was considered to be a subspecies of *O. laciniata* (Munz 1965; Dietrich 1977), but based on recent analysis of crossing relationships was placed in the new subsect. *Nutantigemma* (Dietrich & Wagner 1987). Crossing experiments (Stubbe & Raven 1979) showed that hybrids between *O. pubescens* and species of subsect. *Raimannia* are sterile and that chromosome pairing in meiosis is irregular. The most obvious morphological feature distinguishing *O. pubescens* and the other two species grouped with it from the members of subsect. *Raimannia* is the nodding flower buds, a characteristic that occurs within sect. *Oenothera* only in *O. pubescens, O. breedlovei,* and *O. tamrae.* Consequently there is a clear morphological and genomic distinction between the species of subsect. *Nutantigemma* and those of the other subsections of sect. *Oenothera.*

Subsection *Raimannia* can be divided into two clearly defined series based on morphological differences in the density of flowers in the inflorescence and the shape of the capsules. The inflorescences of the species of ser. *Raimannia* are lax, frequently interrupted by lateral shoots, the petals are truncate or emarginate, and the fruits are long and cylindrical, whereas the inflorescences of species of ser. *Candela* are dense, uninterrupted spikes, the petals are acute to rounded, and the capsules are relatively shorter and thickened basally. Each of the series forms a distinctive crossing group. In general, hybrids between species of ser. *Candela* are fully fertile and have normally functioning plastids that produce green hybrids. Likewise, hybrids between species of ser. *Raimannia* are usually fully fertile and green. These two series are, however, morphologically more distinct from each other than the species of ser. *Raimannia* are from those of subsect. *Nutantigemma.* They are placed together in subsect. *Raimannia* primarily because the crossing studies showed that their genomes and plastomes are similar and interfertile; however, interseries hybrids produce pale green plants with poorly functioning plastids, but if the hybrids reach the flowering stage, they have a basically normal meiosis with no irregularities and exhibit normal seed-set. The level of differentiation of the series within subsect. *Raimannia* is similar to that found between ser. *Renneria* (Fischer) W. Dietrich and ser. *Allochroa* (Fischer & Meyer) W. Dietrich of the South American subsect.

Munzia. The species of ser. *Candela* probably represent a lineage that diverged relatively early in the evolution of subsect. *Raimannia*. All bivalent-forming species of ser. *Candela* have retained genetic self-incompatibility, whereas of the bivalent-forming species in ser. *Raimannia*, only *O. grandis* is self-incompatible.

The evolution of the two series of subsect. *Raimannia* parallels in many ways the situation for the related subsect. *Munzia*. Both subsections consist of two clearly differentiated, but related, groups. The distributional range of both ser. *Candela* and ser. *Renneria* are more limited, whereas ser. *Raimannia* and ser. *Allochroa* have greater ranges extending to the Atlantic Ocean, and in some cases (e.g., *O. drummondii* subsp. *thalassaphila* and *O. stricta*, respectively) even to the Pacific coast. Both ser. *Candela* and ser. *Renneria* have densely flowered, compact spikes and fruits which are thicker toward the base, whereas in series *Raimannia* and *Allochroa* the inflorescences are fewer-flowered and the fruits are cylindrical. The comparison in the evolution between the North and South American species, however, ends with these observations. The South American species exhibit a much more complex pattern of diversification, including a third group, ser. *Clelandia* W. Dietrich, whose species are all permanent structural heterozygotes, derived via hybridization between species of series *Renneria* and *Allochroa*. In contrast, the permanent structural heterozygotic species of subsect. *Raimannia* all appear to have been formed from within a single taxonomic species. The absence of hybrid species in subsect. *Raimannia* is presumably due to the fact that although hybrids between species of ser. *Candela* and ser. *Raimannia* have fertile seeds, the plastids of hybrid progeny are nonfunctional or function at a low level in the background of the foreign genome of the other series. Thus under normal conditions hybrid combinations between ser. *Candela* and ser. *Raimannia* would not be viable even though there are numerous sympatric occurrences. In subsect. *Munzia*, however, a great many of the species of ser. *Renneria* and ser. *Allochroa* are fully cross-compatible (Dietrich 1977), and therefore, in subsect. *Munzia*, it has been possible for permanent heterozygotic species to arise via hybridization.

Series *Candela* and ser. *Raimannia* represent morphologically, cytologically, and geographically distinct lineages. Only *O. cordata* is somewhat intermediate between the series. It has a more open, fewer-flowered spike than the other species of ser. *Candela*, petals that are more rounded, and capsules that are narrowly lanceoloid but are of a size typical for ser. *Candela*. *Oenothera cordata* thus appears to represent somewhat of a transitional phase from ser. *Candela* to ser. *Raimannia*. It is, however, clearly a member of ser. *Candela*, because it has exactly the same crossing behavior and the same general habit as other species of ser. *Candela*.

DISTRIBUTION AND ECOLOGY

The genus *Oenothera* appears to have originated in northern Mexico and the adjacent United States. This hypothesis is based on the occurrence of all but four of the fourteen currently recognized sections of the genus in this area. Furthermore, the least specialized species, in general, occur within the area and the more specialized species usually occur beyond it. Subsection *Raimannia* is centered within this area; however, ser. *Candela* has spread somewhat less from this presumed center of origin than the overall range of ser. *Raimannia* or subsect. *Nutantigemma*. The considerable extent of the range of ser. *Raimannia* and subsect. *Nutantigemma* is due to just a few species. The range of ser. *Raimannia* extends eastward beyond that of ser. *Candela* with *O. laciniata,* which occurs in nearly all counties in the eastern United States, and terminates at the Atlantic coast, where *O. drummondii* subsp. *drummondii* and *O. humifusa* inhabit coastal dunes. Likewise, *O. pubescens* of subsect. *Nutantigemma* has an extensive range well beyond northern Mexico southward into Andean South America.

The species of ser. *Candela* appear to have diversified ecologically less than the species of ser. *Raimannia.* They occur primarily in grasslands of the central United States, whereas the species of ser. *Raimannia* occur in other habitats, including recently disturbed sites, open sites in forest, and coastal dunes.

Comparison of the distribution of the permanent structural heterozygotic species with that of the bivalent-forming species most closely related to them, shows that the permanent heterozygotes occur in geographical areas much farther from and often ecologically different from the presumed center of origin of subsect. *Raimannia* in Texas. *Oenothera clelandii* is centered in the central lowlands, from eastern Minnesota to northern Indiana and Michigan, whereas *O. rhombipetala,* from which *O. clelandii* presumably was derived, occurs from Texas to Nebraska in the Great Plains region. Obviously, the eco-geographically defined areas have shifted greatly during the Pleistocene and earlier, and the evolution of these groups did not occur, for the most part, under present-day circumstances; nonetheless, the kinds of relationships involved doubtless suggest those that occurred in the past. Autogamous, permanent structural heterozygotes in general possess more narrowly defined arrays of genotypes than do their outcrossing relatives; such genotypes may fit them precisely to particular habitats marginal to their ancestral populations.

The case of *O. curtissii,* which also appears to have originated from *O. rhombipetala,* and which is adapted to more humid localities than its progenitor, seems to parallel that of *O. clelandii.* In addition to this general genetic advantage, the autogamous structural heterozyotes likewise avoid the necessity for obligate outcrossing, which is mandatory in *O. rhombipetala* and may be more difficult to achieve in ecologically marginal areas. The only species of ser. *Candela* to become naturalized outside its native range is one of the permanent structural heterozygotes, *O. clelandii.*

Similar observations can be made for the permanent heterozygotes in ser. *Raimannia. Oenothera laciniata* is much more widespread than its progenitor species *O. grandis;* it occurs in a much greater diversity of habitats and has even become a nearly cosmopolitan weed. In the case of *O. drummondii* and its permanent structural heterozygote derivative, *O. humifusa,* the latter has been able to colonize much farther north along the Atlantic coast than *O. drummondii.* A similar pattern is found in comparing *O. breedlovei* and its permanent structural heterozygote derivative, *O. pubescens* in subsect. *Nutantigemma,* which has the most extensive natural range of any species in the genus; its natural range extends from New Mexico and Arizona south to Guatemala, Andean South America, and the Galapagos Islands. The only exception to the pattern is *O. drummondii* subsp. *drummondii,* which is a bivalent-forming, modally outcrossing species; it has been widely naturalized in arid coastal dune areas in many parts of the world. It is one of the few bivalent forming species of *Oenothera* to become naturalized and the only one in subsect. *Raimannia.*

TAXONOMY

Note: The shape and measurements of "mature buds" in the following descriptions refer only to the portion enclosed by the sepals and exclude the floral tube and ovary.

Oenothera Linnaeus, Sp. pl. 346. 1753.—Lectotype, designated by Rose, 1905: *Oenothera biennis* Linnaeus. (For generic synonymy, see Raven 1964.)

Annual, biennial or perennial herbs, acaulescent or caulescent, with erect, ascending or rarely decumbent stems, when decumbent sometimes rooting at the nodes, with a taproot or fibrous roots, occasionally with adventitious shoots forming from spreading lateral roots or rarely with rhizomes. Basal rosette present or absent in mature plants; leaves alternate, entire, toothed to pinnatifid, often irregularly so, stipules absent. Flowers in the leaf axils, when numerous forming terminal bracteate spikes, racemes or corymbs, actinomorphic, perfect, 4-merous, opening near sunset or near sunrise, usually ephemeral; floral tube well developed, cylindrical, somewhat flared at the mouth, deciduous soon after anthesis; petals yellow, purple, or white, rarely with a red spot or entirely red, usually aging orange or purple, usually obovate or obcordate; stamens 8; anthers versatile, the sporogenous tissue in each locule undivided, pollen shed singly, connected by viscin threads; ovary 4-locular; stigma deeply divided into 4 linear lobes, entire surface receptive. Fruit a capsule, sometimes tardily dehiscent, rarely indeshicent, straight or curved, terete to 4-angled or winged, sessile or the basal portion constricted, sterile, and stipelike. Seeds in 1–2 (–3) rows or in clusters in each locule. Chromosome numbers: $n = 7, 14, 21, 28$.

The genus *Oenothera* comprises 124 species, primarily occurring in open or disturbed habitats from North to South America; several species are widely naturalized. It is divided into 14 sections, 12 of which occur in Texas and northern Mexico. The largest section by far is sect. *Oenothera* with 76 species, subdivided into 5 subsections. Revisionary work is currently in progress on a number of the sections (Wagner, in prep.; Raven et al., in prep.), and therefore a key to sections is not given here.

Oenothera section **Oenothera.**

Annual, biennial or perennial herbs, caulescent or rarely nearly acaulescent, with erect to decumbent stems, from taproots, rarely from fibrous roots or with adventitious shoots forming from spreading lateral roots. Basal rosette present, rarely few-leaved and fugaceous. Flowers usually in dense to lax terminal spikes, sometimes solitary in well-spaced leaf axils, opening near sunset; petals yellow, rarely red or with a red spot at base, obovate, obcordate, broadly elliptic or rhombic-elliptic, occasionally suborbicular. Capsules cylindrical to narrowly lanceoloid or ovoid, 4-angled or terete. Seeds in (1–) 2 rows per locule, prismatic and angled, or ellipsoid to subglobose, rarely obovoid and obtusely angled, the testa reticulate and regularly or irregularly pitted, rarely flat.

Key to the Subsections of Oenothera Sect. Oenothera, and to the Series and Species of Subsect. Raimannia and Subsect. Nutantigemma

1. Plants perennial herbs with a multistemmed shrubby habit or with long decumbent or weakly ascending stems from a basal rosette, and sometimes rooting at the nodes; floral tube 5.5–19 cm long; northern Mexico and adjacent southern New Mexico.
subsect. *Emersonia* (Munz) W. Dietrich, Raven & W. L. Wagner (Dietrich et al. 1985).
1. Plants annual, biennial or short-lived perennial herbs with erect to ascending stems, rarely (Andes, South America) forming matlike clumps of leaves and short stems, never with a shrubby habit or long decumbent or weakly ascending stems, never rooting at the nodes; floral tube 1–5 (–13) cm long.
  2. Seeds prismatic, angled, the surface irregularly pitted; Canada to Central America.
  subsect. *Oenothera.*
  2. Seeds ellipsoid to subglobose, not angled, the surface usually regularly pitted.
    3. Young flower buds with floral tube curved upwards or straight and erect.
    I. subsect. *Raimannia.*
      4. Apex of petals acute to rounded. IA. subsect. *Raimannia* ser. *Candela.*
        5. Petals 1.5–3.5 cm long; stigma well elevated above the anthers at anthesis; pollen ca. 90–100% fertile.
          6. Sepals and floral tube glandular-puberulent and with eglandular spreading hairs or glabrous; mature buds usually overtopping the apex of the spike.
            7. Inflorescence dense; bracts longer than the capsules they subtend; sepals and floral tube with pustulate hairs or glabrous. 1. *O. heterophylla.*
            7. Inflorescence lax; bracts shorter than the capsules they subtend; sepals and floral tube without pustulate hairs but never glabrous. 2. *O. cordata.*
          6. Sepals and floral tube strigillose, sometimes also sparsely glandular-puberulent; mature buds not overtopping the apex of the spike. 3. *O. rhombipetala.*
        5. Petals 0.5–1.7 cm long; stigma surrounded by anthers at anthesis; pollen ca. 50% fertile.
            8. Inflorescence dense, more than 2 flowers per spike opening each day; leaves gray-green; central to northeastern U.S.A. 4. *O. clelandii.*
            8. Inflorescence lax, 1–2 flowers per spike opening each day; leaves bright green; Florida, Georgia, Alabama, and South Carolina. 5. *O curtissii.*
      4. Apex of petals truncate or emarginate.
        9. Young flower buds with floral tube curved upward.
        IB. subsect. *Raimannia* ser. *Raimannia.*
          10. Non-flowering portion of stems stiff, densely strigillose or sometimes also villous; leaves grayish green, densely strigillose, subentire to remotedly shallowly dentate, rarely lyrate, bracts flat; coastal sites.
            11. Sepals 1.3–3.3 cm long; petals 2–4.5 cm long; stigma elevated above the anthers at anthesis; pollen 90–100% fertile. 10. *O. drummondii.*
            11. Sepals 0.3–1.1 cm long; petals 0.45–1.6 cm long; stigma surrounded by the anthers at anthesis; pollen ca. 50% fertile. 11. *O. humifusa.*
          10. Non-flowering portion of stems not stiff, moderately to sparsely strigillose and sparsely to sometimes densely villous, and more or less glandular-puberulent; leaves green, sparsely to moderately strigillose and usually villous, deeply lobed to dentate or rarely some of them subentire, or if grayish green, then the bracts revolute; inland sites, often in disturbed habitats.
            12. Petals 2.5–4 cm long; style 4–7.5 cm long, the part exserted from the floral tube 1.5–3 cm long, stigma lobes well elevated above the anthers at anthesis; pollen 90–100% fertile. 6. *O. grandis.*
            12. Petals 0.5–2.5 cm long; style 2–5 cm long, the part exserted from the floral tube 0.3–2 (–2.5) cm long, stigma lobes surrounded by or slightly elevated above the anthers at anthesis; pollen ca. 50% fertile or 90–100% fertile.

13. Leaves usually grayish green, densely strigillose, margins revolute; uppermost bracts erect. 8. *O. mexicana.*
13. Leaves usually green, sparsely to moderately strigillose and more or less villous, margins not revolute; uppermost bracts spreading.
    14. Stigma lobes usually slightly elevated above the anthers at anthesis, the part of the style exserted from the floral tube 1.2–2.5 cm long; petals 1.3–2.5 cm long; pollen 90–100% fertile. 7. *O. falfurriae.*
    14. Stigma lobes surrounded by the anthers at anthesis, the part of the style exserted from the floral tube 0.3–1.4 cm long; petals 0.5–2.2 cm long; pollen ca. 50% fertile. 9. *O. laciniata.*

9. Young flower buds with floral tube straight. subsect. *Munzia* W. Dietrich (Dietrich 1977).

3. Young flower buds with floral tube recurved, nodding. II. subsect. *Nutantigemma.*
15. Stigma elevated above the anthers at anthesis; pollen ca. 90–100% fertile; Laguna Mts., Baja California, and Nayarit.
    16. Lower leaves deeply parted almost to the midrib; mature buds 3–5 mm in diameter at base; Laguna Mts., Baja California Sur. 12. *O. breedlovei.*
    16. Lower leaves not parted to the midrib; mature buds 5–7 mm in diameter at base; Nayarit. 13. *O. tamrae.*
15. Stigma surrounded by the anthers; pollen ca. 50% fertile; Arizona, Texas, Mexico (except Baja California), Guatemala, Colombia to Peru. 14. *O. pubescens.*

I. Oenothera section **Oenothera** subsection **Raimannia** (Rose ex Britton & A. Brown) W. Dietrich, Ann. Missouri Bot. Gard. 64: 612. 1977 [1978]. *Raimannia* Rose ex Britton & A. Brown, Ill. fl. n. U.S., ed. 2, 2: 596. October 1913. *Oenothera* subg. *Raimannia* (Rose ex Britton & A. Brown) Munz, Amer. J. Bot. 22: 645. 1935. *Oenothera* subg. *Raimannia* sect. *Raimannia* (Rose ex Britton & A. Brown) Munz, N. Amer. Fl. II. 5: 105. 1965.—Type: *Oenothera laciniata* Hill.

Erect to procumbent annual or perennial herbs, sometimes forming a rosette or only with a few basal leaves; stems green or flushed with red, simple to much-branched, strigillose or strigillose and villous, in the region of the inflorescence sometimes also glandular-puberulent or rarely subglabrous. Rosette leaves very narrowly oblanceolate to lanceolate, or very narrowly elliptic to elliptic, pinnately parted to remotely and bluntly dentate, the lobes sometimes dentate, apex acute to almost obtuse, gradually narrowed to the petiole; cauline leaves narrowly oblong to oblong, very narrowly lanceolate to lanceolate, very narrowly elliptic to broadly elliptic or ovate to broadly ovate, lobed to subentire, lobes sometimes dentate, apex acute or rarely almost rounded (*O. drummondii*), base narrowly cuneate to subcordate, sessile or gradually narrowed into a short petiole, rarely abruptly narrowed (*O. drummondii*); all leaves strigillose or villous, sometimes the hairs essentially confined to the margins, rarely also glandular-puberulent. Inflorescence dense to lax, simple or with lateral branches, mature buds not exceeding the apex of the spike, rarely elevated above it, with floral tube straight and erect or curved upward. One to several flowers per spike opening each day near sunset. Floral tube 1.5–5 cm long, yellowish, often flushed with red, sometimes also red-flecked, strigillose or glandular-puberulent, strigillose and villous or villous and glandular-puberulent, sometimes glabrous or long-hirsute, the hairs with pustulate bases (*O. heterophylla* subsp. *heterophylla*). Petals yellow to pale yellow, fading orange to reddish orange or becoming colorless when wilted, very broadly obovate, broadly obovate to orbicular or rhombic, apex acute to rounded or truncate to emarginate. Stigma elevated above the anthers at anthesis, or surrounded by the anthers and pollen shed directly onto the stigma. Capsules narrowly lanceoloid to lanceoloid or cylindrical, straight or somewhat curved upward and in an obtuse angle to the stem. Seeds ellipsoid to subglobose, brown to dark brown, often with darker flecks, 0.8–2 mm long, the surface usually regularly pitted. Self-incompatible or self-compatible. Base chromosome number: x = 7.

Species of subsect. *Raimannia* usually grow at relatively low elevations, unlike the species of subsect. *Nutantigemma,* which only occur at altitudes above 1500 m. The ecological range of subsect. *Raimannia* includes temperate deciduous forests, grasslands of the Great Plains, and dunes of the Atlantic, Pacific, and Gulf of Mexico coasts.

IA. Oenothera section **Oenothera** subsection **Raimannia** series **Candela** W. Dietrich & W. L. Wagner, Ann. Missouri Bot. Gard. 74: 147. 1987.—TYPE: *Oenothera rhombipetala* Nuttall ex Torrey & A. Gray.

Erect annual, biennial, or probably short-lived perennial herbs from a taproot and forming a rosette; stems green or flushed with red, simple or branched only in upper part or from the base with branches arising obliquely from the rosette, strigillose, sometimes also villous or glandular-puberulent or in the region of the inflorescence sometimes glabrate. Rosette leaves very narrowly oblanceolate to oblanceolate or very narrowly elliptic, deeply lobed to subentire; cauline leaves very narrowly lanceolate to elliptic, oblong, ovate to broadly ovate or narrowly oblanceolate, lobed to subentire; all leaves very densely to sparsely strigillose on both surfaces, sometimes glabrate, uppermost bracts occasionally glandular-puberulent. Inflorescence dense to open and lax (*O. cordata*), unbranched or with lateral flowering branches near base of central spike. Mature buds with floral tube straight and erect. Two to several flowers per spike opening each day near sunset. Petals yellow, rarely pale yellow, usually becoming colorless when wilted, broadly elliptic to more or less rhombic or rotund to suborbicular, acute to rounded at apex. Stigma elevated above the anthers at anthesis or surrounded by the anthers and pollen shed directly onto the stigma. Capsule narrowly lanceoloid to lanceoloid, straight or curved upward. Seeds narrowly ellipsoid to broadly ellipsoid, brown to dark brown, often dark-flecked.

Series *Candela* comprises a clearly defined group of five closely related species. They all have the presumably derived character of relatively densely flowered spikes on which two or more flowers open every evening, unlike the species of ser. *Raimannia,* which nearly always produce only one flower per branch each day, a plesiomorphic character. The spikes of ser. *Candela* never have lateral shoots, as is often the case in species of ser. *Raimannia.* In cultivation under normal conditions in Düsseldorf, *O. curtissii* is an exception; as a short-day plant it is reluctant to flower and forms lateral shoots instead of flower buds, a phenomenon that has not been observed in nature. The fully grown buds of ser. *Candela* are narrowly oblong to narrowly lanceoloid, whereas those of ser. *Raimannia* are lanceoloid in the large-flowered, structural homozygotic species and oblong to ellipsoid in the small-flowered, heterozygotic species. In ser. *Raimannia,* the older buds are curved upwards until shortly before flowering, whereas those of ser. *Candela* are straight. The shape of the petals also clearly differentiates the species of ser. *Candela* from those of ser. *Raimannia;* those of ser. *Candela* are acute to rounded at the apex, whereas those of ser. *Raimannia* are truncate to emarginate. Furthermore, the capsules of ser. *Candela* are relatively short and thicker towards the base, whereas those of ser. *Raimannia* are on the average longer and nearly cylindrical. All species of ser. *Candela* appear to be biennials. In contrast to this specialized habit, the species of ser. *Raimannia* have evolved an annual habit or in the case of *O. drummondii* and *O. humifusa,* which inhabit sand dunes, have retained the generalized perennial habit.

Both series of subsect. *Raimannia* appear to have originated in the southeastern portion of Texas or in surrounding areas in grasslands or savannas that resembled the present Gulf prairies, Blackland prairies, South Texas plains, or post-oak savanna of southeastern Texas. Both series are best represented in terms of numbers of species in these areas. Moreover, the species with plesiomorphic characters, such as large flowers and self-incompatibility, as well as the derived, narrow endemic species like *O. mexicana* and *O. falfurriae* occur here.

The cytological studies by Hecht (1950) and Cleland (1968), as well as an intensive analysis at the Botanical Institute of the University of Düsseldorf, showed that the chromosomes of *O. heterophylla, O. cordata,* and *O. rhombipetala* are structurally homozygotic and therefore form 7_{II} or sometimes one to three rings of chromosomes in meiotic metaphase I. *Oenothera clelandii* and *O. curtissii* are permanent structurally heterozygotic species which form a ring of 14 chromosomes in meiotic metaphase I; approximately 50% of their pollen is aborted. With the exception of the strain "Alabama" of *O. heterophylla* subsp. *orientalis,* the three homozygotic, large-flowered species are self-incompatible.

1. Oenothera heterophylla Spach, Nouv. Ann. Mus. Hist. Nat. 4: 348. 1835. *Raimannia heterophylla* (Spach) Rose ex Sprague & Riley, Bull. Misc. Inform. 1921: 200. 1921. *Oenothera variifolia* Steudel, Nomencl. Bot. ed. 2, 2: 208. 1841, nom. superfl.—TYPE: U.S.A., Texas, II [second North American trip], 1833–34, *Drummond 74* (holotype: FI!; isotypes: E! G! GH! K! P!).

Erect annual or short-lived perennial herbs, forming a rosette with many or sometimes with only a few leaves; stems 2.5–7 dm long (in cultivation up to 16 dm long), up to ca. 8 mm in diameter, green or flushed with red, simple or branched mainly in upper part, sparsely to densely strigillose and in the inflorescence sometimes nearly glabrous. Rosette leaves 7–15 cm long, 1–2.5 cm wide, very narrowly oblanceolate to oblanceolate, deeply lobed to subentire, apex acute to almost obtuse, gradually narrowed to the petiole; cauline leaves 3–13 cm long, 0.4–2.3 cm wide, very narrowly lanceolate to lanceolate or very narrowly elliptic to elliptic, lobed to remotely dentate, apex acute, base rounded to narrowly cuneate, sessile or short-petiolate; bracts 1–3 cm long, 0.3–1.2 cm wide, narrowly lanceolate to narrowly ovate or ovate, remotely dentate to subentire, apex acute to acuminate, base narrowly cuneate to subcordate, sessile; leaves and bracts sparsely strigillose or the surfaces sometimes nearly glabrous. Inflorescence dense, often with several lateral spikes just below the main one. Two to several flowers per spike opening each day near sunset. Floral tube 2.5–4.7 cm long, ca. 1 mm in diameter, sometimes flecked with red, sparsely glandular-puberulent, sometimes also sparsely strigillose, villous, or glabrous, often with additional long spreading hairs, each of the long hairs arising from a long several-celled pustule. Mature buds 3–5 mm in diameter at the base, narrowly oblong or narrowly lanceoloid, usually overtopping the apex of the spike. Sepals 1.5–3 cm long, yellowish, sometimes flecked with red, pubescence like that of floral tube, sepal tips 1–6 mm long, spreading to erect in bud, strigillose or glabrous. Petals 2.5–3.5 cm long, 1.5–2.5 cm wide, yellow, broadly elliptic to nearly rhombic, apex acute. Filaments 1.5–3 cm long; anthers 3–8 mm long, pollen ca. 90–100% fertile. Ovary 0.4–0.7 cm long, ca. 1 mm in diameter, densely to sparsely strigillose, sometimes also sparsely glandular-puberulent, or sometimes nearly glabrous. Style 4.5–7.5 cm long, the visible part 1.8–3.5 cm long; stigma usually elevated above the anthers at anthesis, the lobes 2–5 mm long. Capsule 1.3–2.5 cm long, 2.5–4 mm in diameter at base, lanceoloid, usually curved upward, pubescence like that of ovary. Seeds 1.1–1.8 mm long, 0.4–0.8 mm in diameter, ellipsoid to broadly ellipsoid, sometimes obscurely angled, brown, often flecked with darker spots, the surface pitted. Self-compatible or self-incompatible, modally outcrossing.

Oenothera heterophylla is distinguished from other species in ser. *Candela* by its relatively short, thick spikes with the mature buds and flowers elevated above its apex. In robust plants, short lateral spikes often arise from the base of the main spike. In cultivation, *O. heterophylla* forms a rosette of only a few leaves; by contrast, *O. rhombipetala* and *O. clelandii* have rosettes with many leaves. Moreover, it forms a main stem soon after germination, whereas *O. rhombipetala* and *O. clelandii* remain much longer in the rosette stage. This suggests that in its natural habitat *O. heterophylla* usually grows as an annual, whereas the two other species are typically biennials.

Oenothera heterophylla subsp. *orientalis* is distinguished from subsp. *heterophylla* primarily by its lower growth habit as well as by the absence of pustulate-based hairs, which are characteristic for subsp. *heterophylla.* The calyx lobes are on the average shorter and the sepals never have red coloration. Among the strains studied at the Botanical Institute in Düsseldorf, the one from Arkansas was self-incompatible, like subsp. *heterophylla,* whereas the strain from Alabama was self-compatible. In cultivation, subsp. *orientalis* flowers earlier in the season than subsp. *heterophylla.* Of all the strains of ser. *Candela* that have been cultivated, the Alabama strain of *O. heterophylla* subsp. *orientalis* flowered the earliest.

Key to the Subspecies of Oenothera heterophylla

1. Floral tube and sepals always with pustulate-based hairs and often with red spots; sepal tips 2–6 mm long, usually spreading in bud; ovary and capsule never glabrous. 1a. *O. heterophylla* subsp. *heterophylla.*
1. Floral tube and sepals without pustulate-based hairs and red spots; sepal tips 1–3 mm long, erect and appressed in bud; ovary and capsule often glabrous. 1b. *O. heterophylla* subsp. *orientalis.*

1a. Oenothera heterophylla subsp. **heterophylla.** *Oenothera heterophylla* var. *heterophylla,* Fosberg, Amer. Midl. Naturalist 27: 763. 1942.

Oenothera pyramidalis var. *lindheimeri* H. Léveillé, Monogr. Onothera 383. 1909.—Type: U.S.A. Texas: [probably Waller or Austin Co.], prairie in eastern Brazos [River] bottom, 1843, *Lindheimer 56* (lectotype, here designated: MO-2529699!; isolectotypes: BM! FI! GH! K! MIN! PH!). Locality reconstructed with aid of McKelvey (1955: 901). Léveillé cites three collections, all of which are *Oenothera heterophylla* subsp. *heterophylla;* we have selected the one most widely represented in herbaria.

Rosette leaves usually few, parted up to ⅔ the width to subentire; base of bracts truncate to subcordate. Inflorescence often with several secondary spikes. Floral tube 2.5–4.2 cm long, sometimes flecked with red, always with pustulate hairs. Sepals 1.5–2.8 cm long, sometimes flecked with red, pubescent with pustulate-based hairs, sepal tips 2–6 mm long, usually spreading. Petals 1.8–3.5 cm long, 2–3 cm wide. Ovary and capsule densely to sparsely strigillose and glandular-puberulent. Self-incompatible. Chromosome number: $n = 7$ (7_{II}*, ⊙4 and 5_{II}**, or 2 ⊙4 and 3_{II}*** at meiotic metaphase I). Fig. 1.

Phenology. Flowering from June to September.

Distribution (Fig. 2). Eastern Texas to eastern Louisiana.

Specimens from Cultivated Plants. U.S.A. Texas: Nacogdoches Co., Nacogdoches, S end of University Drive, 1974, *Hoff s.n.* [cult. no. 75–1406*, 75–1424***, 75–1426**, 75–1427*, 75–1430*, 75–1431*, 75–1432*, 75–1434*] (DUSS, M, MO).

Representative Specimens. U.S.A. Louisiana: Caddo Pa., Rogers Station, ca. 5 mi S of Vivian, *Thieret 31220* (DS, GA, LL). Calcasieu Pa., 2.5 mi SE of Moss Bluff, *Thieret 31638* (DS, GA). Erwin Pa., Rte. 156 at Saline Bayon, *Cooke 1277* (ARIZ). Natchitoches Pa., Goldonna, *Dormon s.n.* (BH, SMU). Winn Pa., 1.75 mi SE of Goldonna, *Thieret 30305* (DAO, GA, LL, LSU).—Missouri: St. Louis Co., St. Louis, along tracks of Terminal RR Assoc., N of O'Fallon St., *Mühlenbach 1018* (MO).—Texas: Anderson Co., 17.75 mi S of Palestine, *Cory 21934* (POM). San Augustine Co., 1958, *Smith s.n.* (DS). Bastrop Co., 1.75 mi W of Bastrop, *Cory 51673* (NY). Brazos Co., College Station, 1946, *Parks s.n.* (RSA, TAES). Cass Co., 5 mi N of Linden, *Whitehouse 21369* (SMU). Chambers Co., E of Anahuac, *Tharp & Gimbredge 52631* (COLO, OKLA, TEX). Cherokee Co., 3.1 mi N of Farm Road 241 off Hwy 21, 1958, *Brannon s.n.* (ASTC). Dallas Co., Dallas, *Reverchon 295* (US). Freestone Co., 15 mi S of Teague, *Harding 396* (OKLA). Gonzales Co., Gonzales, *Trécul 1438D* (P). Gregg Co., Gladewater, *Shinners 24073* (SMU). Hardin Co., Village Creek, 7.5 mi W of Silsbee, *Cory 11270* (POM, TAES). Harris Co., Simslake, *Boon 182* (GH, MO, NY, TEX, UC). Harrison Co., 1 mi S of Marshall on Hwy 59, *Nixon et al. s.n.* (TEX). Henderson Co., 1 mi W of Athens, *Sanders 156* (ARIZ, MICH, SMU). Hopkins Co., 3 mi E of Greenville, 1950, *Stitteler s.n.* (PH). Houston Co., Grapeland, *Tharp 860* (TEX, US). Jasper Co., Jasper, *Clark 3506* (G). Lee Co., SW of Old Dime Box, *Fryxell 2510* (NY). Leon Co., mouth of Buffalobayon in Galveston Bay, 1842, *Lindheimer s.n.* (MO). Liberty Co., Dayton, *Munz 13344* (POM). Limestone Co., near Navasota River Bridge, 7 mi SW of Mexia, 1959, *Sewell s.n.* (ASTC). Nacogdoches Co., Nacogdoches, *Waller 154* (DS, ILL, TAES, TEX, UC). Newton Co., Sabine River bottom E of Call, *Knight 86* (SMU, TEX). Robertson Co., 8 mi S of Hearne on Hwy 6, *Walker 17* (TENN, UARK, WVA). Rusk Co., 3.5 mi N of Mount Enterprise, *Griffin 1* (ASTC). Sabine Co., 4 mi N of Hemphill, *Toole 11* (ASTC). San Jacinto Co., Evergreen, *Fisher 14* (POM). Smith Co., Swan, *Reverchon 3561* (MO, NY, US). Sutton Co., Underpass, *Reeves 178* (TAES). Travis Co., Bejar near Austin, 1842, *Berlandier s.n.* (GH). Tyler Co., 17 mi S of Woodville, *Cory 49957* (MICH, SMU). Upshur Co., 4.4 mi SE of Gilmer, *Shinners 31843* (CM, DAO, FLAS, G, GA, GH, KE, SMU, TENN, UNA, WIS, WVA). Van Zandt Co., near Conton, *Bass 10* (SMU). Victoria Co., 10 mi SW of Victoria, *McVaugh 12379* (MICH, RSA, US). Wood Co., Mineola, *Reverchon 245* (MO). County unknown: *Lindheimer 35* (MIN); 1848, *Wright s.n.* (GH); 1843, *Lindley s.n.* (K).

The most striking characteristic of *O. heterophylla* subsp. *heterophylla* is the unusual pubescence on the sepals and the floral tube. This pubescence consists of stout hairs 3–4 mm long, which grow out of a narrow, multicellular base up to 1 mm long. Hairs with a pustulate base, a characteristic that usually has been referred to

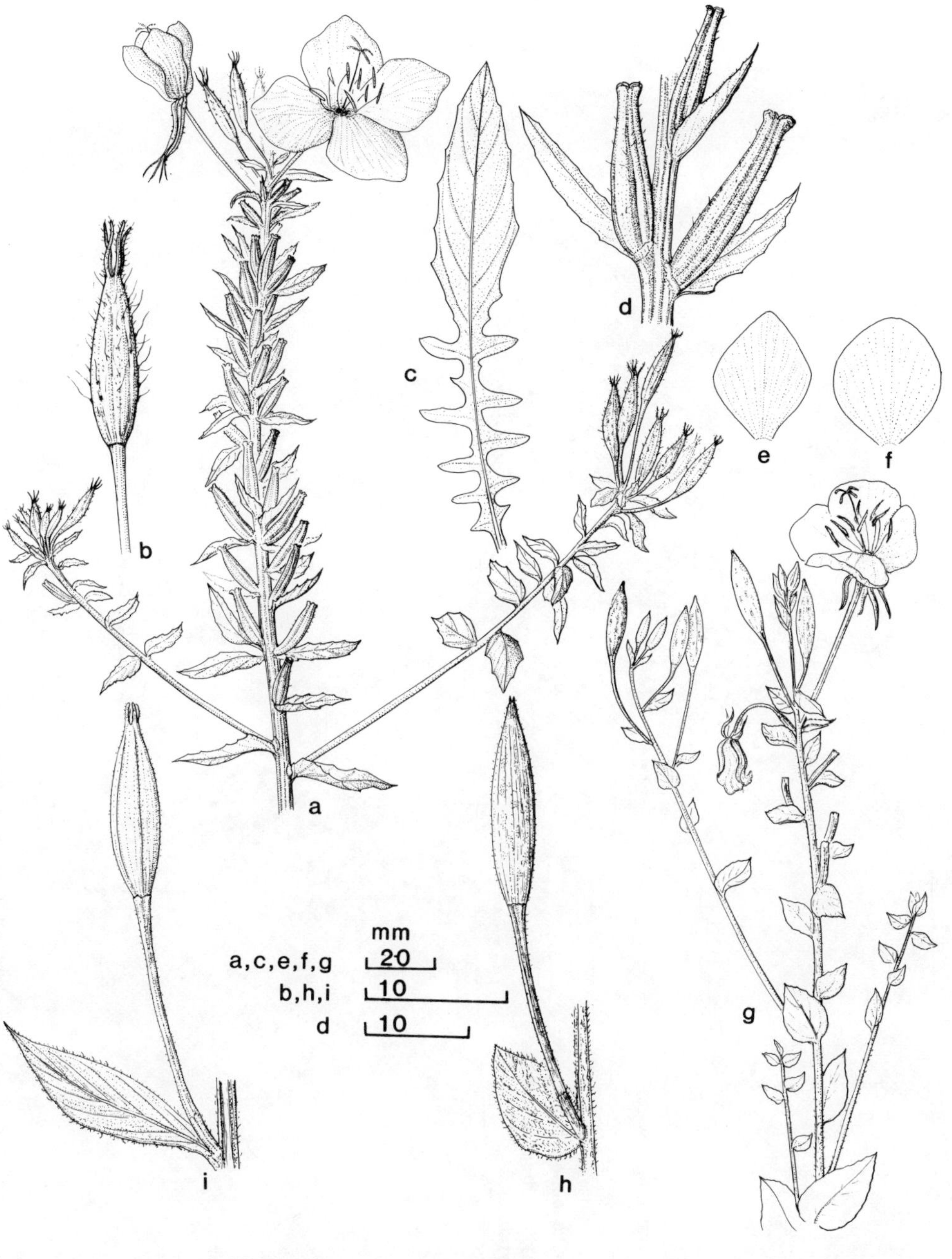

FIG. 1. *Oenothera heterophylla* and *O. cordata. O. heterophylla* subsp. *heterophylla* (*Hoff s.n.* in 1974, cult. no. 76–715, 76–717; MO): a. Inflorescence. b. Bud. c. Leaf (76–715). d. Capsules. e. Petal. *O. cordata* (*Ellstrand s.n.* in 1978, cult. no. 79–95; MO): f. Petal. g. Inflorescence. h. Bud with bract. *O. heterophylla* subsp. *orientalis* (*Demaree 70094,* cult. no. 76–706; MO): i. Bud with bract.

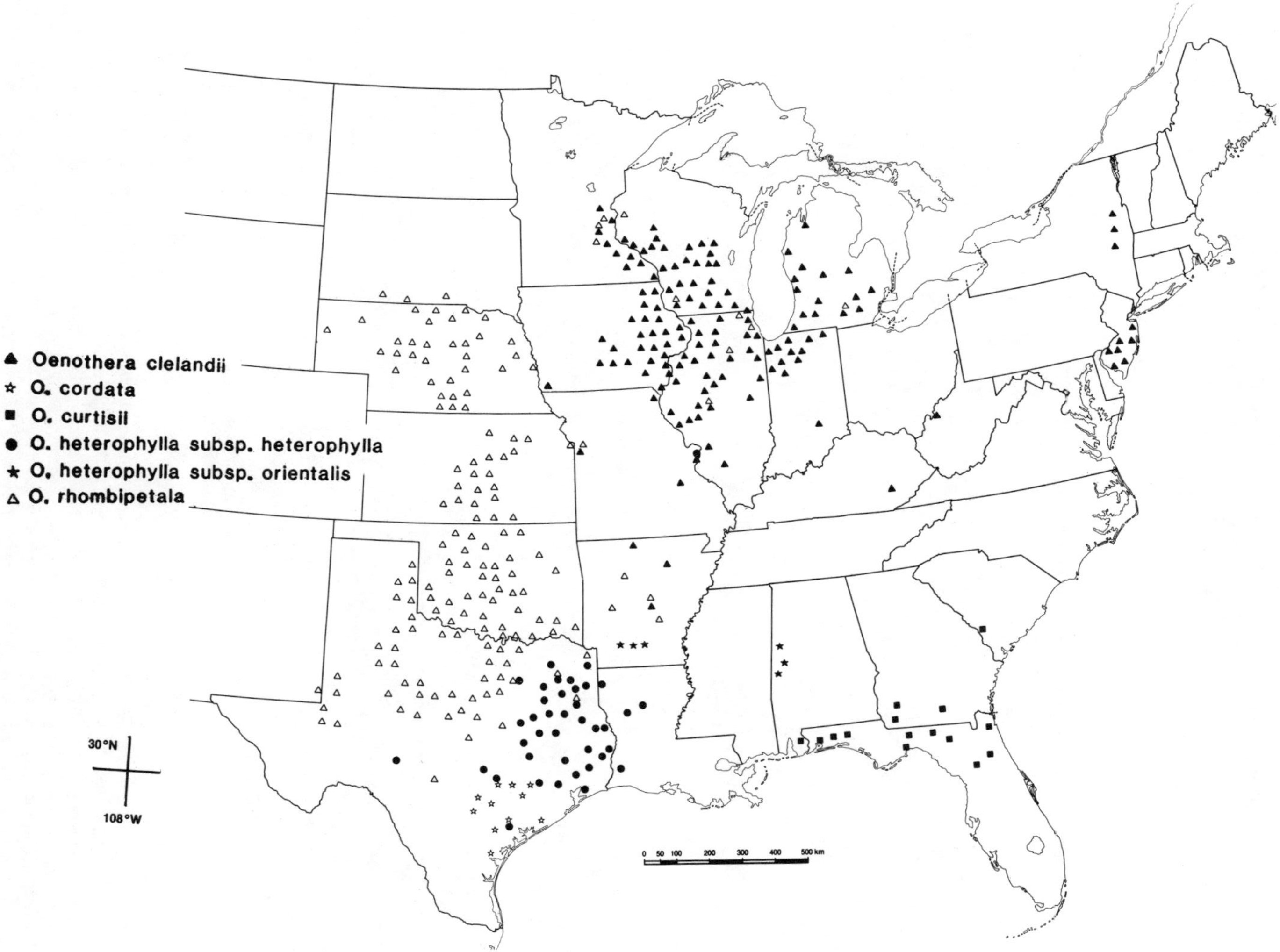

FIG. 2. Distribution of species of *Oenothera* sect. *Oenothera* subsect. *Raimannia* ser. *Candela: Oenothera cordata, O. clelandii, O. curtissii, O. heterophylla* subsp. *heterophylla, O. heterophylla* subsp. *orientalis,* and *O. rhombipetala.*

in the literature as "muricate," is common in *Oenothera,* and is especially striking in the species of subsect. *Oenothera* (for instance *O. elata* H. B. K.). The bases of pustulate hairs in species other than *O. heterophylla,* however, are much shorter and are always wider than they are long. They also are usually red or purple, whereas those of *O. heterophylla* are generally colorless.

Oenothera heterophylla subsp. *heterophylla* is self-incompatible. As in many other species of *Oenothera,* a number of different small translocation rings (floating translocations) occur within populations of subsp. *heterophylla.* Similar situations occur in such species as those of *Oenothera* subsect. *Munzia* of South America (Dietrich 1977), subsect. *Emersonia* (Dietrich et al. 1985), as well as in many strains cultivated in Düsseldorf of *O. elata* [including *O. hookeri* Torrey & A. Gray] in subsect. *Oenothera* and in many other species of the tribe Onagreae. Structurally different chromosomal homozygotes occur within *O. heterophylla* subsp. *heterophylla,* and hybrids between them form small rings of chromosomes. Unlike *O. rhombipetala,* which has given rise to the autogamous, permanent structural heterozygotic species *O. clelandii* and *O. curtissii, O. heterophylla* has not produced any permanent structural heterozygote derivatives, although it has several chromosomal end arrangements.

Morphological variation also can be conspicuous within populations of *Oenothera heterophylla* subsp. *heterophylla.* For example, variation in the Nacogdoches strain ranges from plants with deeply divided rosette and lower stem leaves to those with only serrate leaf margins. The sepals and floral tubes of these plants are either uniformly light yellow or have red spots.

1b. Oenothera heterophylla subsp. **orientalis** W. Dietrich, Raven & W. L. Wagner, in W. L. Wagner, Ann. Missouri Bot. Gard. 70: 196. 1983.—TYPE: U.S.A. Alabama: Sumter Co., sandy clay field, 5.2 mi S of Dancy, 3 May 1972, *Kral 46366* (holotype: MO-2379457!; isotypes: DUKE! GA! GH! NCU! NY! USF! VDB!).

Rosette leaves usually numerous, deeply parted from ⅔ the width to the midrib; cauline leaves deeply parted to subentire; base of bracts narrowly cuneate to cuneate. Inflorescence usually simple. Floral tube 3–4.7 cm long, without pustulate-based hairs. Sepals 1.7–3 cm long, without pustulate-based hairs, sepal tips 1–3 mm long, erect. Petals 2.5–3.5 cm long, 1.5–2.5 cm wide. Ovary and capsule often glabrous, sometimes strigillose and glandular-puberulent. Self-compatible or self-incompatible. Chromosome number: $n = 7$ (7_{II}* at meiotic metaphase I). Fig. 1.

Phenology. Flowering from May to July.

Distribution (Fig. 2). Known from two disjunct areas: Greene, Pickens, and Sumter counties, Alabama, and Calhoun, Nevada, and Ouachita counties, Arkansas. *Oenothera heterophylla* subsp. *orientalis* is sparsely represented in herbaria, and its actual range might be more extensive than understood at present.

SPECIMENS FROM CULTIVATED PLANTS. U.S.A. ALABAMA: Pickens Co., 4.3 mi S of Aliceville, 1974, *Kral s.n.* [cult. no. 75–1436*] (DUSS, M, MO).—ARKANSAS: Nevada Co., Bluff City, 1975, *Demaree 70094* [cult. no. 76–706*] (DUSS, M, MO).

ADDITIONAL SPECIMENS EXAMINED. U.S.A. ALABAMA: Greene Co., W of Smith Lake, *Harper 4522* (GH, NCU); Boligee along Hwy 11, *Kral 12109* (VDB); Tombigbee River, by RR crossing N of Epes and S of Boligee, just N of Hwy 11, *Kral 50093* (GA, NY, US, VDB). Pickens Co., 4 mi S of Aliceville on Hwy 17, *Kral 45316* (VDB), *Kral 46243* (AUA, DUKE, GH, M, NCU, NY, US, USF, VDB); 4.3 mi S of Aliceville, *Kral 46777* (MO, NCU, US, VDB); S of Aliceville along Tombigbee River, *Rogers 9199* (NCU). Sumter Co., 5.2 mi S of Dancy, *Kral 46366* (DUKE, GA, GH, MO, NCU, USF, VDB). County unknown: Big-Ben-River, *Mohr 524* (US).—ARKANSAS: Calhoun Co., Tinsman, *Demaree 22664* (DS). Nevada Co., Bragg Lake near Bluff City, *Demaree 48158* (DS, DUKE, FLAS, FSU, G, GA, NA, NY, UC, WVA); Bluff City, *Demaree 52542* (MO), *Demaree 69749, Demaree & Graham 655* (MASS, NCU, VDB), *Moore 56-134* (UARK); Bragg City, *Demaree 55987* (MASS). Ouachita Co., Chidester, *Moore 56-114* (APCR, F, NY); White Oak Lake near Chidester, *Tucker 5731* (MASS, NCU), *Tucker 8079* (NCU).

SPECIMENS INTERMEDIATE BETWEEN *Oenothera heterophylla* SUBSP. *heterophylla* AND SUBSP. *orientalis*. U.S.A. TEXAS: Brazos Co., 10 mi SE of College Station, *Celarier 51–92* (OKL); College Station, 1946, *Parks s.n.* (TAES). Limestone Co., *Tharp 2917* (TEX). Smith Co., 3 mi S of Lindale, 1961, *Lackey s.n.* (ASTC).

2. Oenothera cordata J. W. Loudon, Ladies' flower-gard. 1: 167. 1840 [date according to Stafleu and Cowan (1981), reissued in 1843]. *Oenothera bifrons* D. Don in Sweet, Brit. Fl. Gard. ser. 2, 4: *t. 386.* 1838, non *Oenothera bifrons* Lindley, Edwards's Bot. Reg. 17: *t. 1405.* 1831.—TYPE: No authentic material seen. The illustration attached to Don's description, which is based on plants grown from seeds collected by Drummond in Texas during his second North American trip, is here designated as the lectotype.

Erect annual or biennial herbs; stems 2.5–7 dm long, ca. 0.4–0.6 cm in diameter, green or flushed with red, simple or branched above, densely to sparsely strigillose, sometimes also sparsely villous and glandular-puberulent in the upper parts. Rosette leaves 6–12 cm long, 0.7–2 cm wide, very narrowly elliptic to oblanceolate, remotely and bluntly dentate or lobed, apex acute, gradually narrowed to the petiole; cauline leaves 2–10 cm long, 0.5–3 cm wide, narrowly lanceolate, narrowly oblong, narrowly elliptic, or ovate to broadly ovate, remotely and bluntly dentate, lobed, or subentire, apex acute, base narrowly cuneate to subcordate, short-petiolate to sessile; bracts 0.5–1.7 cm long, 0.4–1.3 cm wide, narrowly ovate to broadly ovate, denticulate to subentire, apex acute, base rounded to subcordate, sessile, shorter than the capsule they subtend; leaves and bracts densely to sparsely strigillose, bracts also sometimes glandular-puberulent. Inflorescence rather lax, simple, usually with only a few flowers per spike. One to two flowers per spike opening each day near sunset. Floral tube 2–4 cm long, ca. 1.5 mm in diameter, yellowish, often flushed with red or with red spots, sparsely villous and densely to sparsely glandular-puberulent or exclusively glandular-puberulent. Mature buds 3–5 mm in diameter, narrowly oblong in outline, usually overtopping the apex of the spike. Sepals 1.5–2.5 cm long, greenish to yellowish, often flushed with red or with red spots, pubescence like that of floral tube, sepal tips 0.3–1 mm long, erect in bud, strigillose to villous. Petals 2–3 cm long, 1.5–2.7 cm wide, yellow, rotund to suborbicular, apex rounded to obtuse. Filaments 1.7–2.2 cm long; anthers 4–7 mm long, pollen ca. 90–100% fertile. Ovary 0.5–1 cm long, ca. 1 mm in diameter, densely strigillose, sometimes also glandular-puberulent. Style 5–6.5 cm long, the exserted part 2–3 cm long; stigma elevated above the anthers at anthesis, the lobes 2.5–5 mm long. Capsule 1.5–3.3 cm long, 2–3 mm in diameter, very narrowly lanceoloid. Seeds 1–1.4 mm long, 0.4–0.6 mm in diameter, ellipsoid, dark brown, the surface pitted. Self-incompatible. Chromosome number: $n = 7$ (7_{II}*, or ⊙4 and 5_{II}** at meiotic metaphase I). Fig. 1.

Phenology. Flowering from April to July.

Distribution (Fig. 2). Sandy open places in southeastern Texas.

SPECIMENS FROM CULTIVATED PLANTS. U.S.A. TEXAS: Guadalupe Co., Seguin on Hwy 123, 1978, *Ellstrand s.n.* [cult. no. 79–97*, 79–98*, 79–99*] (DUSS, M, MO). Wilson Co., Stockdale on Hwy 123, 1978, *Ellstrand s.n.* [cult. no. 79–94**, 79–95*, 79–96*] (DUSS, M, MO).

ADDITIONAL SPECIMENS EXAMINED. U.S.A. TEXAS: Austin Co., 1931, *Tharp s.n.* (TEX); between Sealy and Belville on Hwy 36, *Jones 53* (M). Bastrop Co., near Willis Mileys, *Duval 165* (TEX). Colorado Co., Columbus, 1910, *Rusby s.n.* (NY). Fayette Co., Colony, *Crawford 8* (MO), *Crawford 32* (US). Goliad Co., Goliad, *Williams 169* (TEX). Guadalupe Co., 10 mi N of Stockdale on Hwy 123, 1964, *Raven & Gregory s.n.* (RSA); 10 mi S of Seguin, *Webster & Rowell 7091* (TEX). Matagorda Co., Bay City, *Palmer 9626* (DS, US). San Patricio Co., 1 mi S of Ingleside, *Jones 481* (SMU). Victoria Co., Da Costa School, *Faifer 7214* (TEX, US). Waller Co., Hempstead, *Hall 201* (BM, F, K, MO, NA, NY, POM, US). Wilson Co., Kicaster School, *Cory 15104* (POM); Terrell Hill, *Cory 15105* (POM); 3 mi N of Floresville, *Munz 13333* (BH, CS); Terrell Hill near Kicaster, *Munz 13335* (BH, GH, POM); Kicaster School, *Parks & Cory 15101, 15103* (TAES), *Parks & Cory 1502* (SMU); 5 mi N of Stockdale, *Turner 4964* (TEX). County unknown: *Eman-Meyer s.n.* (NY); "Texas oriental," 1848–49, *Wright s.n.* (G); Texas II [second trip], *Drummond 53* (K); 1889, *Hopkins s.n.* (US).

SPECIMENS INTERMEDIATE BETWEEN *Oenothera heterophylla* SUBSP. *heterophylla* AND *O. cordata.* U.S.A. TEXAS: Bastrop Co., near Bastrop, 1932, *Albers s.n.* (TEX). Tyler Co., 17 mi S of Woodville, *Cory 49957* (SMU). Victoria Co., ca. 10 mi SW of Victoria between hwy & RR, *McVaugh 12379* (MICH, RSA, US).

Oenothera cordata has usually been considered to be conspecific with *O. heterophylla;* however, this self-incompatible species is distinguished from *O. heterophylla,* to which it is closely related, by its open lax inflorescence and broad subcordate bracts that are distinctly shorter than the fruits. The bracts at the tip of the spike cover the young buds, whereas in *O. heterophylla* they are spreading, exposing the young buds. The petals of *O. cordata* are rounded at the apex, a characteristic that is not shared with any other species of ser. *Candela.* Petals in all other species of ser. *Candela* are always acuminate. They also differ from the petals of species of ser. *Raimannia,* which are truncate or emarginate.

The earliest known collection of *O. cordata* was made in the 1830's (*Drummond 53*). David Don described the species in 1838 on the basis of plants from the "Bristol Nursery." These plants were grown by a Mr. Miller who had received seeds from Drummond. Even though we have not yet seen any authentic herbarium material, we were able to identify the collections here assigned to *O. cordata* as being the same as the plant illustrated by Don. The distinctive characters are the few-flowered inflorescence, the broad subcordate bracts, the upper bracts erect and covering the young buds, and the rounded petals. The name *O. cordata* was applied to a number of collections from the mid-1800's, all of which are actually *O. heterophylla* subsp. *heterophylla.*

3. Oenothera rhombipetala Nuttall ex Torrey & A. Gray, N. Amer. Fl. 1: 493. 1840. *Raimannia rhombipetala* (Nuttall ex Torrey & A. Gray) Rose ex Britton & A. Brown, Ill. fl. n. U.S., ed. 2, 2: 597. 1913. *Oenothera pyramidalis* H. Léveillé, Monogr. Onoth. 382. 1909, nom. superfl. *Oenothera heterophylla* var. *rhombipetala* (Nuttall ex Torrey & A. Gray) Fosberg, Amer. Midl. Naturalist 27: 763. 1942.—TYPE: [U.S.A. Oklahoma: Chatow Co.], plains of Red River, Jun 1819, *Nuttall s.n.* (lectotype, here designated: BM!; isolectotypes: K! US!).

Oenothera leona Buckley, Proc. Acad. Nat. Sci. Philadelphia 1861 [1862]: 455. 1861.—TYPE: U.S.A. Texas: near Leon River, Jun 1861, *Buckley s.n.* (lectotype, here designated: PH-0949826!, pro parte; isolectotype: GH!). A specimen of *Oenothera heterophylla* subsp. *heterophylla* is also mounted on the lectotype sheet.

Erect biennial herbs from taproots, forming rosettes; stems 3–10 (–15) dm long, up to 1.5 cm in diameter at base, usually green, simple or branched, sometimes with lateral branches arising obliquely from the rosette, densely to sparsely strigillose, sometimes also glandular-puberulent in the upper parts. Rosette leaves 6–20 cm long, 0.6–2 cm wide, narrowly oblanceolate, bluntly dentate to lobed, apex acute, gradually narrowed to the petiole; cauline leaves 3–15 cm long, 0.8–2.5 cm wide, narrowly elliptic to narrowly lanceolate or narrowly oblanceolate to ovate, subentire or dentate to lobed, apex acute, base narrowly cuneate to truncate, sessile or short-petiolate; bracts 0.8–2.5 cm long, 0.4–2 cm wide, narrowly lanceolate to narrowly ovate, entire or dentate, apex acute, base obtuse to rounded, sessile, shorter or slightly longer than the capsule they subtend; leaves and bracts usually densely strigillose. Inflorescence dense, without lateral branches. Two or more flowers per spike opening each day near sunset. Floral tube 3–4.5 cm long, 1–1.5 mm in diameter, densely to sparsely strigillose and sometimes also glandular-puberulent. Mature buds 3–6 mm in diameter at base, narrowly lanceoloid to narrowly oblong, not overtopping the apex of the spike. Sepals 1.5–3 cm long, yellowish, sometimes flecked with red, pubescence like that of floral tube, sepal tips 0.5–3 mm long, erect in bud, strigillose to almost glabrous. Petals 1.5–3.5 cm long, 1.2–3 cm wide, yellow, broadly elliptic to rhombic-elliptic, apex acute. Filaments 1.3–2.3 cm long; anthers 3–8 mm long, pollen ca. 90–100% fertile. Ovary 5–8 mm long, 1–1.5 mm in diameter, densely strigillose and sometimes also glandular-puberulent. Style 2.5–5 cm long; stigma elevated above the anthers at anthesis, lobes 2–5 mm long. Capsule 1.3–2.5 cm long, 2.5–3 mm in diameter, narrowly lanceoloid, strigillose and sometimes also sparsely glandular-puberulent. Seeds 1–1.7 mm long, 0.4–0.7 mm in diameter, ellipsoid, brown, sometimes flecked with dark red spots, the surface pitted. Self-incompatible. Chromosome number: $n = 7$ (7_{II}*, ⊙4 and 5_{II}**, or 3 ⊙4 and 1_{II}*** at meiotic metaphase I). Fig. 3.

Phenology. Flowering from May to October.

Distribution (Fig. 2). Fields and prairies from southern South Dakota, Nebraska, Kansas, western Missouri, Oklahoma, Texas, scattered areas in central Arkansas, and Lea County, New Mexico; populations in eastern Minnesota, southwestern Wisconsin, Illinois, and Michigan probably represent introductions or perhaps relicts.

SPECIMENS FROM CULTIVATED PLANTS. U.S.A. KANSAS: Dickinson Co., 5 mi E of Solomon, *Raven 26528* [cult. no. 77–515*] (DUSS, M, MO); 1 mi N of Sand Springs, *McGregor 26327* [cult. no. 74–1417**] (DUSS, M, MO). Kingman Co., near Kingman, *Brooks 8619* [cult. no. 75–1416**] (DUSS, M, MO). Reno Co., Cheney Reservoir, *Brooks 8620* [cult. no. 75–1419**] (DUSS, M, MO). Summer Co., near Mayfield, *Brooks 8638* [cult. no. 75–1418*] (DUSS, M, MO).—MINNESOTA: Anoka Co., Fridley, 1974, *Ownbey s.n.* [cult. no. 76–701**] (DUSS, M, MO).—OKLAHOMA: Blaine Co., 2 mi E of Bridgeport, 1974, *Goodman s.n.* [cult. no. 75–1409***] (DUSS, M, MO). Canadian Co., Methodist Camp in Devils Canyon, 1973, *Estes s.n.* [cult. no. 75–1402**] (DUSS, M, MO).—TEXAS: Collingsworth Co., 2.5

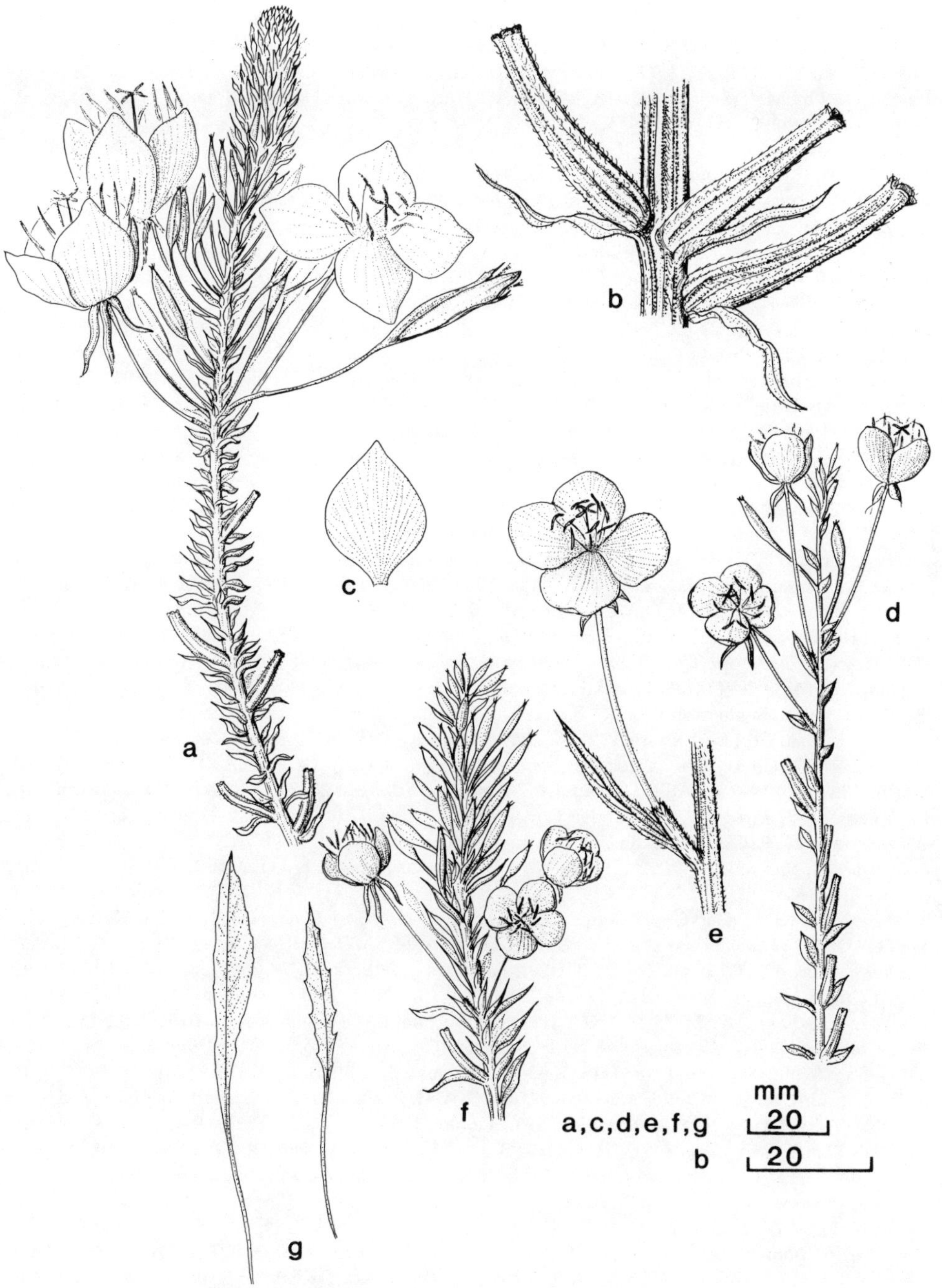

FIG. 3. *Oenothera rhombipetala, O. curtissii,* and *O. clelandii. O. rhombipetala* (*Stephens 81703,* cult. no. 76–705; MO): a. Inflorescence. b. Capsules. c. Petal. *O. curtissii* (*Godfrey 76646;* MO): d. Inflorescence. *O. clelandii* (*Barkley s.n. in 1974,* cult. no. 78–205; MO): e. Flower with bract. f. Inflorescence. g. Rosette leaves.

mi W of Quail, *Stephens 80966* [cult. no. 75–1422**] (DUSS, M, MO). Garza Co., 2.2 mi N of Brazos River, *Benbow 82* [cult. no. 77–509*] (DUSS, M, MO). Wheeler Co., 2.5 mi E of Kellerville, *Stephens 81703* [cult. no. 75–1421*] (DUSS, M, MO). Winkler Co., 10 mi NE of Kermit, *Kolle 1413* [cult. no. 77–510*, 82–711**] (DUSS, M, MO).——FRANCE: Botanical Garden of Conches, seeds from North America collected by M. Trécul in 1851 (K).—U.S.A.: Hort. Cantabr. (Harvard Botanical Garden), seeds from Texas, herb. A. Gray in 1844 (K).

REPRESENTATIVE SPECIMENS. U.S.A. ARKANSAS: Jefferson Co., Yell Bend, *Stern 274* (UARK). Montgomery Co., Washita, *Leavenworth s.n.* (GH). Pope Co., Holla Bend Wildlife Refuge, *Tucker 7207* (NCU). Pulaski Co., Natural Steps, *Demaree 8740* (GH, MO, NY, SMU).—ILLINOIS: Cook Co., Chicago, *Vasey s.n.* (MASS). La Salle Co., Starved Rock, 1884, *Waite s.n.* (DUKE). McHenry Co., Ringwood, 1866, *Vasey s.n.* (BM). Menard Co., Athens, 1863, *Hall s.n.* (F), 1864, 1865, 1866, *Hall s.n.* (P), 1867, *Hall s.n.* (FI).—KANSAS: Barton Co., Ellenwood, *Rydberg & Imler 1323* (KSC, MO, NY). Cloud Co., *Fraser 165* (KSC). Cowley Co., Winfield, *Koch 1807* (OKLA). Dickinson Co., Abilene, *Jackson 177* (F, KANU, NCU, NY, SMU). Ellsworth Co., Kanopolis, 1896, *Becker s.n.* (KSC). Geary Co., Junction City, *Brooks 8130* (MO). Harper Co., *Hitchcock 687* (GH, KSC, MO, NY, UNM, US). Harvey Co., Burrton, *Stephens 19154* (DS, KANU). Kingman Co., Kingman, *Bare 2500* (KANU). Kiowa Co., Belvidere, 1891, *White s.n.* (KSC). McPherson Co., 1940, *Gustafson s.n.* (KSC). Pottawatomie Co., 1904, *Beigman s.n.* (NDA). Pratt Co., Pratt, *Stephens 87695* (KANU). Reno Co., Hutchinson, *McGregor 15850* (KANU, NCU, SMU). Rice Co., Raymond, *Stephens 34587* (KANU). Riley Co., 1921, *Dickens s.n.* (KSC). Sabine Co., Niles, *Raven & Gregory 19489* (DS). Saline Co., Hedville, *Brooks 8129* (KANU). Sedgewick Co., Wichita, *Bartley 393* (US). Stafford Co., Hudson, *McGregor 12489* (KANU, US). Sumner Co., Mayfield, *Brooks 8638* (KANU, MO). Wyandotte Co., Turner, 1895, *Mackenzie s.n.* (NY).—MICHIGAN: Jackson Co., Grass Lake Twp, *Parmelee 764* (MSC).—MINNESOTA: Anoka Co., Cesar Creek Natural History Area, 1959, *Ovington s.n.* (MIN). Hennepin Co., Minneapolis, 1891, *Arton s.n.* (DAO, ILL, MIN, MSC, MU, NY, YU). Scott Co., 1938, *McCartney s.n.* (MIN).—MISSOURI: Jackson Co., Sheffield, *Bush 820* (K).—NEBRASKA: Adams Co., Holstein, *Stephens 18997* (DS, KANU). Antelope Co., Elgen, *Stephens & Brooks 24183* (DS, KANU). Arthur Co., Arthur, *Stephens & Brooks 24914* (DS, KANU). Blaine Co., Brewster, *Stephens & Brooks 24351* (DS, KANU). Boyd Co., Butte, *Stephens 15473* (DS, KANU). Brown Co., 31 mi S of Ainsworth, *Stephens 24418* (DS, KANU). Buffalo Co., Fort Kearney, 1856, *Engelmann s.n.* (GH, MO). Butler Co., Bellwood, *Stephens 60548* (KANU). Cherry Co., Valentine, *Stephens & Brooks 34216* (ASC, KANU, SD). Cuming Co., Pilger, *Churchill 1994* (MASS). Custer Co., Anselms, *Stephens 28175* (DS, KANU). Dawson Co., Lexington, 1924, *Eifrig s.n.* (F). Dodge Co., Fremont, *Kiener 21497* (F, IA). Douglas Co., Elk City, *Wiegand et al. 1719* (CU). Franklin Co., Franklin, *McGregor & Bare 1724* (KANU). Garfield Co., 16 mi N of Burwell, *Iltis 18523* (WIS). Greeley Co., Greeley, *Stephens 62118* (KANU). Harlan Co., Alma, *Brooks 8128* (KANU, NCU). Holt Co., Atkinson, *Stephens 15516* (DS, KANU). Hooker Co., Mullen, *Stephens 17166* (DS, KANU). Howard Co., Dannebrog, *Stephens 15768* (DS, KANU). Kearney Co., Minden, 1933, *Hapeman s.n.* (ARIZ, CM, ILLS, MASS, SMU, TENN). Keya Paha Co., Norden, *Stephens & Brooks 34293* (KANU, VDB). Lincoln Co., Sutherland, *Munz 17531* (BH, NY, POM). Logan Co., Stapleton, *Stephens & Brooks 25033* (DS, KANU). McPherson Co., 22 mi W of Tryon, *Stephens & Brooks 24998* (DS, KANU). Phelps Co., Funk, *McGregor & Bare 1678* (KANU). Platte Co., Columbus, 1904, *Clevinger s.n.* (US). Sheridan Co., Spade Ranch, 1905, *Buchanan s.n.* (ISC). Sioux Co., Agate, *Stephens 16183* (DS, KANU). Thomas Co., Plummer Ford, *Rydberg 1510* (C, GH, NY). Webster Co., Red Cloud, *McGregor 25624* (KANU). Wheeler Co., Bartlett, *Stephens & Brooks 24265* (DS, KANU).—NEW MEXICO: Lea Co., Hobbs, 1941, *Cory s.n.* (POM).—OKLAHOMA: Alfalfa Co., Salt Plains, *Goodman 2160* (GH, ISC, MO, OKL). Atoka Co., Tushka, *Cory 56820* (SMU, USF). Beckham Co., Elk City, *Munz & Gregory 23511* (RSA, UC). Blaine Co., Canton Reservoir, *Kelting 249* (ILL, KANU, OKL, RSA, UC). Bryan Co., Colbert, *Cory 52849* (SMU). Caddo Co., Ft. Cobb, *Demaree 13062* (AC, GH, MIN, MO, NY, OKL, PH, POM, SMU, TEX, US). Canadian Co., Ft. Reno, 1896, *Blankenship s.n.* (GH, US). Carter Co., Ardmore, *Kelting 200* (KANU, OKL, UC). Choctaw Co., Fort Towson, 1950, *Nelson & Holland s.n.* (OKL, OKLA, SMU). Cleveland Co., Norman, *Hopkins et al. 790* (MO, OKL, OKLA, UC). Comanche Co., Elgen, 1963, *Seneca s.n.* (NCSC). Custer Co., Weatherford, *Nelson 379* (SIU, SMS). Dewey Co., Taloga, *Palmer 41956* (GH, MO, UC). Garfield Co., Enid, *Hixson 140* (OKLA). Grady Co., Chickasha, *Pearce 784* (KSC, MIN, OKL, SMU). Greer Co., Granite, *Stevens 1035* (DS, GH, ILL, MIN, MO, NY, OKL, OKLA, US). Jackson Co., 2.5 mi W of jct. Hwys 283 & 5, *Barber 920* (OKL, OKLA). Jefferson Co., Urral, *Pryor 273* (OKL). Kay Co., Ponca City, *Davy 54* (OKLA). Kingfisher Co., Kingfisher, *Bollenbach 125* (OKL, TAES). Kiowa

Co., Lake Lugart, *Smith 1026* (OKL, UC). Lincoln Co., Meeker, *Gibson 44* (OKLA). Logan Co., Crescent, *Grace 361* (OKL, POM). Love Co., Marietta, *Williams 357* (NCU, OKL). Major Co., Orienta, *Raven & Gregory 19474* (DS). Marshall Co., Lake Texoma, 1955, *Riggs s.n.* (ILL, OKL, UC). McClain Co., Purcell, *Hopkins & Demaree 104* (ARIZ, GA, IND, OKL, OKLA, P). McCurtain Co., Idabel, *Parkey 77* (OKL). Muskogee Co., Ft. Gibson, *Wallis 7450* (KANU, NC). Oklahoma Co., Oklahoma City, *Waterfall 1474* (BH, OKL). Osage Co., Ponca City, *Bragg 79* (OKL). Payne Co., Ripley, *Long 97* (OKLA). Pontotoc Co., Ada, *McCoy 796* (OKLA). Pottawatomie Co., Tecumseh, 1941, *Rose s.n.* (OKLA). Pushmataha Co., Antlers, *Palmer 8325* (K, MO, NY, US). Seminole Co., Konawa, *Robbins 3087* (NY, SMU, TAES, UC). Stephens Co., Comanche, *Lipscomb 726* (UARK). Tillman Co., Frederick, *Edwards 129* (OKLA). Tulsa Co., Sand Springs bridge, *Clark 360* (OKL, OKLA). Woods Co., Greensburg, *Nighswonger 869* (NCU, OKL). Woodward Co., *Smith 526* (OKL).—South Dakota: Bennett Co., Tuthill, *McGregor 25634* (KANU). Gregory Co., Burke, *Stephens 49242* (KANU, NCU). Todd Co., Little White River, *Over 16134* (US).—Texas: Andrews Co., 10 mi S of Andrews, *Vandergriff 31* (TTC). Bowie Co., DeKalb, *McCrary 81* (TEX). Brown Co., 16 mi N of Brownwood, *Chamberlain 9* (SMU, TEX). Callahan Co., Clyde, *Henderson 63-816* (ISC, SMS, SMU, TEX, VT). Childress Co., Memphis, *Higgins 7243* (NY). Coke Co., Silver, *Powell 16012* (MO). Collingsworth Co., Quail, *Stephens 80966* (KANU, MO). Comanche Co., Proctor, *Grimes 261* (SMU, TENN). Cooke Co., Gainesville, *Harding 490* (OKLA). Cottle Co., Paducah, *Stephens 80675* (KANU, MO). Crane Co., Crane, *Warnock 15425* (TEX). Crosby Co., Crosbyton, 1968, *Morris s.n.* (NY). Dallas Co., Dallas, *Lundell 13905* (TEX, UC, US). Denton Co., Denton, *McCart 1102* (SMU). Dickens Co., Dickens, *Proder 43* (OKL, OKLA). Donley Co., Hedley, *Stephens 81021* (KANU, MO). Eastland Co., 12 mi S of Asco, *Gould 7586* (SMU, TAES, TEX). Erath Co., Stephenville, *Hoisington 319* (ILL, OKL, RSA). Fisher Co., Roby, *Cory 32202* (POM, TAES). Gaines Co., 15 mi W of Lamesa, *Whitehouse 16778* (MICH, SMU). Garza Co., Post, *Hutchins 694* (OKL, SMU, TEX, TTC). Grayson Co., Denison, 1950, *Reed s.n.* (SMU). Hemphill Co., Canadian, *Rowell 4260* (TEX, TTC). Hood Co., 15 mi NE of Granbury, *Reed 1258* (NCSC, OKL, SMU). Johnson Co., Alvarado, *Shinners 10201* (SMU). Jones Co., Anson, *Cory 15805* (POM). Kent Co., Jayton, *Lundell 13054* (TEX). Kerr Co., Kerrville, *Reed 3286* (TTC, US). Lampasas Co., Lampasas, 1943, *Plummer & Bartley s.n.* (F, GH, MO, NY, TEX, UC). McLennan Co., W of Marlin Hwy, *Smith 594* (OKLA, TEX). Mitchell Co., Colorado City, *Pohl 4965* (ISC, PENN, SMU). Montague Co., Nocona, *Whitehouse 10062* (MICH, SMU). Motley Co., Matador, *Higgins 6110* (NCU). Parker Co., Weatherford, *Tracy 7820* (BM, E, F, G, GH, MO, MSC, TEX, US). Runnells Co., Ballinger, *Palmer 10334* (DS, MO, PH, US). Rusk Co., 1902, *Böckeler s.n.* (BREM). Tarrant Co., Lake Worth, *Ruth 493* (CU, ILL, KSC, MICH, NY, PAM, PENN, PH, POM, SMU, TENN, TEX, US, WIS). Taylor Co., Abilene, *Tolstead 7485* (MICH, MO, POM, SMU, TEX, UC). Ward Co., Monahans, *Muller 8518* (MICH, NY, OKLA, SMU, TEX). Wheeler Co., Kellerville, *Stephens 81703* (KANU, MO). Wichita Co., 10 mi N of Electra, *Whitehouse 10478* (MICH, NY, SMU). Wilbarger Co., Vernon, 1903, *Reverchon s.n.* (US). Winkler Co., Kermit, *Irving 93* (SMU, TAES, TEX). Wise Co., Springtown, *Shinners 26431* (FSU, G, SMU, WIS). Wood Co., Mineola, *Reverchon 2002* (US).— Wisconsin: Grant Co., 1861, *Hale s.n.* (MIN). Polk Co., La Crosse, 1861, *Hale s.n.* (MO, US, WIS).

Specimen from Outside Natural Area. Afghanistan: Mazarisharif, 1937, *Koelz 13192* (NA).

Oenothera rhombipetala, which is primarily distributed in the Great Plains region of the United States, is distinctive, together with *O. clelandii,* in its elongated, densely flowered spikes. At least during the earlier portion of the flowering season, the rachis is elevated well above the open flowers and mature buds. This arrangement contrasts sharply with the inflorescence of *O. heterophylla* and *O. cordata,* in which the flowers extend above the growing apex of the spike. *Oenothera rhombipetala* is closely related to *O. clelandii* and differs primarily in being a self-incompatible bivalent former with 90–100% fertile pollen. Related to these features are the large flowers, in which the stigma is clearly elevated above the anthers. The cytological studies suggest a similar pattern of variation in terminal arrangement of the chromosomes, the presence of a ring of 4 chromosomes in some plants, as observed in *O. heterophylla.*

4. Oenothera clelandii W. Dietrich, Raven & W. L. Wagner in W. L. Wagner, Ann. Missouri Bot. Gard. 70: 196. 1983.—TYPE: U.S.A. Michigan: Mason Co., Amber Twp, E half of sec. 23 NE ¼, 30 Sep 1974, *Barkley s.n.* (holotype: MO-2383779!; isotype: MO!).

Erect biennial herbs from taproots, forming rosettes; stems 2–7 (–10) dm long, usually green, simple or branched, sometimes with lateral branches arising obliquely from the rosette, densely strigillose. Rosette leaves 5–16 cm long, 0.5–1.5 cm wide, narrowly oblanceolate, remotely short-dentate to deeply lobed, apex acute, gradually narrowed to the petiole; cauline leaves 2–12 cm long, 0.5–2 cm wide, very narrowly elliptic to narrowly lanceolate, the lower ones dentate to lobed, the upper dentate to subentire, apex acute, short-petiolate to sessile, base narrowly cuneate; bracts 1–3.5 cm long, 0.3–1.2 cm wide, narrowly lanceolate to lanceolate, bluntly dentate to subentire, apex acute, base attenuate to truncate; leaves and bracts gray-green, densely to sparsely strigillose. Inflorescence dense, without lateral branches. Two to several flowers per spike opening each day near sunset. Floral tube 1.5–4 cm long, ca. 0.7 mm in diameter, densely to sparsely strigillose, or also sparsely glandular-puberulent. Mature buds 2–3 mm in diameter, narrowly oblong, not overtopping the apex of the spike. Sepals 0.6–1.3 cm long, yellowish, seldom red-flecked, pubescence like that of floral tube, sepal tips 0.5–2 mm long, erect, strigillose. Petals 0.5–1.6 cm long, 0.3–1.1 cm wide, yellow to pale yellow, broadly elliptic to ovate, apex acute, occasionally rounded. Filaments 0.4–1.8 cm long; anthers 2–3.5 mm long, shedding pollen directly on the stigma at anthesis, pollen ca. 50% fertile. Ovary 4–8 mm long, ca. 1 mm in diameter, densely strigillose. Style 2–4 cm long, the exserted part 0.6–1 cm long; stigma surrounded by the anthers at anthesis, lobes 1.5–4 mm long. Capsule 1–2 cm long, 2–3 mm in diameter, narrowly lanceoloid, curved upward or straight, strigillose. Seeds 1–1.9 mm long, 0.4–0.8 mm in diameter, ellipsoid, brown, sometimes flecked with dark brownish red spots, the surface pitted. Autogamous, permanent structural heterozygote. Chromosome number: $n = 7$ (⊙ 14* chromosomes at meiotic metaphase I). Fig. 3.

Phenology. Flowering from June to August, rarely in September.

Distribution (Fig. 2). Fields and prairies in sandy soil in eastern Minnesota, Wisconsin, eastern Iowa, Illinois, northern Indiana, and Michigan; localities in western Missouri, Arkansas, Kentucky, western Virginia, eastern New York, New Jersey, and Ontario, Canada, may represent introductions, although D. Brunton (pers. comm., 1985) suggests that the Ontario locality appears to represent a natural occurrence in a relict prairie site.

SPECIMENS FROM CULTIVATED PLANTS. U.S.A. ILLINOIS: Will Co., near Braidwood, 1974, *Schulenberg & Wilhelm s.n.* [cult. no. 75–1412*] (DUSS, M, MO).—IOWA: Black Hawk Co., 1974, *Allbaugh s.n.* [cult. no. 76–702*] (DUSS, M, MO).—MINNESOTA: Houston Co., Island 19 at Mississippi River mile 694.6, *Ziegler & Leykom 1916* [cult. no. 79–8*] (DUSS, M, MO).—MICHIGAN: Mason Co., Amber Twp, 1974, *Barkley s.n.* [cult. no. 75–1411*] (DUSS, M, MO); near Walhalla, 1974, *Barkley s.n.* [cult. no. 77–503*, 78–205] (DUSS, M, MO); Woodville, 1970, *Steiner s.n.* [cult. no. 79–10*] (DUSS, M, MO).—WISCONSIN: Jefferson Co., 3 mi S of Palmyra, 1973, *Pohl s.n.* [cult. no. 81–110*] (DUSS, M, MO). Portage Co., Stevens Point, 1973, *Freckman s.n.* [cult. no. 75–1403*] (DUSS, M, MO).

REPRESENTATIVE SPECIMENS. CANADA. ONTARIO: Peel Co., ½ km W of Lornewood Creek, 10 m S of C.N. Railway, Lorne Park praire, Missisauga, 43°32′N, 79°37′W, *Brunton & Varga 5399* (MO).—U.S.A. ARKANSAS: Independence Co., Polk Bayou, *Thorrow 9054* (CM, DS, MASS). Marion Co.,

Cotter, *Palmer 5991* (CU, F, MO, P, POM). Pulaski Co., Little Rock, *Merrill 739* (FSU).—ILLINOIS: Adams Co., Quincy, *Wehmeyer 79* (GA, MICH). Bureau Co., New Beford, *Evers 91619* (ILLS). Carroll Co., Thompson, *Ahles 73229* (MASS). Cass Co., Beardstown, *Chase 11334* (ILL). Champaign Co., Urbana, 1948, *Norwood s.n.* (OKL). Cook Co., Chicago, Hyde Park, *Chase 1174* (GH, ILL, MO, NMC, PH, US, VT). Du Page Co., Chicago, 1874, *Munroe s.n.* (K). Fulton Co., Canton, *McDougall 178* (ILL). Henderson Co., Oquauka, 1873, *Patterson s.n.* (F, MICH). Henry Co., Hooppole, *Evers 39629* (ILLS). Iroquois Co., Beaverville, *Evers 18451* (ILLS). Jo Daviess Co., Hanover, *Steyermark 40803* (F). Kankakee Co., St. Anne, *Jones 16650* (ILL). Lake Co., Zion, 1949, *Brown s.n.* (ILL). La Salle Co., Naplale, *Evers 113862* (ILLS). Lee Co., Amboy, *Long 479* (ILL). Madison Co., Poag Station, 1878, *Eggert s.n.* (MIN, MO, SMU). Mason Co., Bath, *Chase 10015* (F, ILL, KANU, NY, OKL, SMU). Menard Co., Athens, 1867, *Hall s.n.* (ISC, NY). Mercer Co., Keithsburg, *Palmer 43687* (MO). Morgan Co., Meredosia, *Evers 130861* (ILLS). Ogle Co., Oregon, *Ahles 4549* (ILL). Pike Co., East Hannibal, 1916, *Davis s.n.* (ILL). Rock Island Co., Rock Island, 1900, *McDonald s.n.* (F, ILL). St. Clair Co., East St. Louis, 1878, *Eggert s.n.* (MO). Scott Co., Winchester, *Evers 10064* (ILLS). Tazewell Co., *Chase 13522* (DAO, ILL, KANU, KSC, MASS, SMU, TEX, WIS). Whiteside Co., Fulton, *Ahles 4257* (ILL). Will Co., Custer Park, *Swink 602* (F). Winnebago Co., Belvidere, *Fell f.45540* (ILL). Woodford Co., Spring Bay, *Chase 10103* (F, ILL, TEX).—INDIANA: Batholomew Co., Mineral Springs, 1961, *Bartlett s.n.* (CU). Elkhart Co., Bristol, *Deam 21000* (GH, IND, NY). Jasper Co., Goodland, *Welch 1844* (MIN). Lagrange Co., Ontario, *Weatherwax 3667* (IND). Lake Co., Millers, 1897, *Umbach s.n.* (F, GH, MICH, MIN, MU, NY, PH, SMU, US). La Porte Co., Michigan City, *Deam 31439* (IND). Marion Co., Ravenswood, 1887, *Lloyd s.n.* (F). Marshall Co., Burr Oak, *Deam 38742* (IND). Newton Co., Enos, *Friesner 22380* (KANU, MICH, OKL, OKLA, SMU, TEX, UC, WVS). Porter Co., Chesterton, 1925, *Churchill s.n.* (AC, CU, MO, US). Starke Co., Knox, *Deam 39551* (IND). White Co., Buffalo, *Deam 39356* (IND, PH).—IOWA: Allamakee Co., Union City Twp, upper Iowa River Valley, *Hayden 10090* (GH, ISC). Benton Co., *Eilers 2591* (IA). Black Hawk Co., Waterloo, 1927, *Pammel s.n.* (ISC, PH, UC). Buchanan Co., Walker, *Eilers 5063* (IA). Cedar Co., Rochester, *Fay 786* (IA). Clayton Co., Marquette, *Fassett 5270* (WIS). Clinton Co., Liberty Twp, *Cooperrider 1813* (IA, ILL). Delaware Co., along Maquoketa River, *Eilers 1778* (IA, RSA). Des Moines Co., Burlington, *Lammers 400* (IA, RSA). Fayette Co., *Fink 452* (US). Fremont Co., Hamburg, 1882, *Hitchcock s.n.* (IA). Iowa Co., Homestead, *Easterly 732* (ARIZ, DAO, IA). Jackson Co., Bellvue, *Cooperrider 1609* (IA, MIN, NCU). Jasper Co., Newton, *Bruggen 725* (IA, UC). Jefferson Co., Fairfield, *Gilly 659* (IA). Johnson Co., Cedar Twp, *Thorne 13655* (IA, UC). Jones Co., Hale Twp, *Cooperrider 1956* (IA). Lee Co., Wever, *Palmer 40566* (MO, PH). Linn Co., Rock Island Rwy Preserve, *Eilers 2227* (IA, RSA). Louisa Co., Columbus Junction, 1903, *Shimek s.n.* (ASC). Mahaska Co., Eddyville, *Augustine 435* (ISC). Marion Co., Red Rock, 1920, *Pammel s.n.* (ISC). Muscatine Co., Muscatine, *Pammel & Reppert 1219* (GH, ISC, MO). Story Co., Ames, 1924, *Pammel & Featherly s.n.* (ISC). Warren Co., 1913, *Pammel s.n.* (ISC). Washington Co., Brighton, *Wagenknecht 639* (KANU). Winneshiek Co., Hesper, 1932, *Shimek s.n.* (ISC).—KENTUCKY: Clay Co., Newfound, *Lloyd s.n.* (DS).—MICHIGAN: Barry Co., Yankee Springs State Game Area, *Eickwort 53* (MSC). Berrien Co., New Buffalo, *Deam 17986* (MICH, MIN). Grand Traverse Co., Traverse City, *Dieterle 1851* (CM, MICH). Jackson Co., Clear Lake, *Hermann 9087* (MICH). Kalamazoo Co., Eagle Lake, *Hermann 9035* (MO, NY). Livingston Co., Edwin S. George Reserve, *Hamerstrom 253* (MICH). Mason Co., Walhalla, 1974, *Barclay s.n.* (MO). Montcalm Co., Greenville, *Voss 7569* (MICH). Muskegon Co., Whitehall, *Voss 2856* (MICH, SMU). Newaygo Co., Newaygo, *Pierce 314* (WIS). Oakland Co., Commerce Twp, *Churchill 73-709* (MSC, NA, VDB). Ottawa Co., N of Pigeon Lake, *Bazuin 6073* (F). Saginaw Co., Geddes, 1956, *Case s.n.* (MICH). Van Buren Co., Decatur, *Luteyn 2060* (DUKE). Washtenaw Co., Sugar Loaf Lake, Lyndon Twp, *Hermann 9140* (MICH).—MINNESOTA: Dakota Co., Castle Rock, 1932, *Rosendahl & Butters s.n.* (MIN). Goodhue Co., Cannon Falls, *Chandler 1603* (MO). Hennepin Co., Minneapolis, 1893, *Sheldon s.n.* (E, NY, US). Houston Co., Jefferson, 1899, *Lyon s.n.* (MIN). Isanti Co., Cedar Creek Natural History Area, *Moore 25669* (MIN). Olmstead Co., Rochester, *Uttal 9424* (MASS, NCU, VPI). Wabasha Co., Lake City, 1886, *Manning s.n.* (MIN). Washington Co., 1 mi S of Hwy 61 from jct. of Co. Road 19 & Hwy 61, *Lindayen 193* (IA, MASS, MIN, NCU, USF). Winona Co., Winona, 1888, *Holzinger s.n.* (MSC).—MISSOURI: Clark Co., Wayland, *Drouit 1787* (GH, UMO). Jackson Co., Sheffield, *Bush 245* (COCO, KSC, MIN, US, VT). St. Louis Co., St. Louis, 1878, *Eggert s.n.* (PRC), *Mühlenbach 942* (MO). Washington Co., Floyd, *Bush 9414* (F, ILL, IND, K, MO, PH).—NEW JERSEY: Atlantic Co., Hammonton, 1937, *Freeman s.n.* (NA). Burlington Co., Green Bank, *Long 26641* (GH, PH, UC). Camden Co., Berlin, *Long 23582* (GH, MIN, PH, UNM). Cumberland Co., Willow Grove, *Long 48786*

(PH). Gloucester Co., Mantua, *Long 49085* (FSU, GH, IND, NY, PH). Monmouth Co., Cotts Neck, *Beals & Bassett 7622* (GH, NY, PH). Ocean Co., Forked River, 1922, *Mackenzie s.n.* (NY, PH).—NEW YORK: Albany Co., Karnes, *House 6425* (NY). Saratoga Co., Clifton Park, *House 22820* (BH, CU, GH, NY, PENN, PRC, UC, US, VPI). Warren Co., Queensberry, *House 27995* (CU, GH, MIN, MO, NY, OKL, PENN, UC).—WISCONSIN: Adams Co., Old Bend of Glacial Lake, *Sorenson 1681* (IA, WIS). Buffalo Co., Fountain City, 1964, *Mickelson s.n.* (WIS). Chippewa Co., Chippewa Falls, *Goessl 1881* (MIL). Columbia Co., Caledonia Twp, 1945, *Allen s.n.* (WIS). Crawford Co., Prairie du Chien, *Smith 7462* (B, MIL, WIS). Dane Co., Middleton, *Fassett 3460* (ISC, WIS). Eau Claire Co., Eau Claire, *Goessl 1611* (MIL). Grant Co., on Mississippi River near Kieler, *Iltis 9467* (UC, WIS). Green Co., Albany, *Fell 58-524* (WIS). Green Lake Co., Princeton, 1968, *Soberalske s.n.* (WIS). Iowa Co., Arena, *Mason 1352* (ARIZ, OKLA, UC, WIS). Jackson Co., Millston, *Jones 40835* (ILL). Jefferson Co., Blue Spring, *Wadmond 3281* (MIN). Juneau Co., Lemonweir, 1956, *Curtis & Greene s.n.* (WIS). Kenosha Co., Kenosha, *Swink 2319* (F). La Crosse Co., La Crosse, *Hartley 1131* (NCU, RSA). Marquette Co., Endeavor, *Rill 1659* (WIS). Monroe Co., Sparta, *Iltis 16949* (SMU). Pepin Co., Pepin, 1884, *Manning s.n.* (MIN). Pierce Co., Prescott, *Museum Expedition 23174* (MIL). Portage Co., Bancroft, *Rill 4095* (WIS). Richland Co., Lone Rock, *Heggelund 5* (WIS). Rock Co., Avon Twp, *Fell 57-675* (WIS). Sauk Co., near Wisconsin River, *Jotter 3878* (MICH), *Smith 8275* (B). Trempealeau Co., Galesville, *Hartley 1714* (USF, WIS). Walworth Co., La Grange, 1941, *Thomson s.n.* (WIS). Waupaca Co., Dayton Twp (T21N, R11E, W½ of SW¼ of Sec. 24), *Underwood 267* (OSH). Waushara Co., Coloma, *Bartlett & Richards 954* (DAO, MICH). Wood Co., W of Nekoosa, *Skroch PE112* (WIS).

The strain referred to by Cleland (1968) as "*heterophylla-rhombipetala*" of unknown origin, which formed a ring of 14 chromosomes in metaphase I of meiosis, represents *O. clelandii.* The "Walhalla" strain mentioned by Hecht (1950) also represents this species.

Except for a few scattered, presumably relictual populations, the range of *O. rhombipetala* does not overlap with that of *O. clelandii.* Where they do occur in the same geographical area, they are not known to occur in mixed populations.

The populations that we divide into three species—*O. rhombipetala, O. clelandii,* and *O. curtissii*—previously were considered to be a single species (Munz 1965). Evidence gathered during this study has shown that the geographically separated small-flowered plants, here treated as *O. clelandii* and *O. curtissii,* are both permanent structural heterozygotes, which, to judge from their distributions and morphological differences, presumably were derived independently from *O. rhombipetala.* In chromosome studies on individuals from eight wild collection sites of *O. clelandii,* all plants formed a ring of 14 chromosomes at meiotic metaphase I. Both complexes are derived from *O. rhombipetala,* as demonstrated by crossing experiments. For example, crosses between *O. rhombipetala* "Texas, Kermit, *Kolle 1413*" (7_{II}) as pistillate parent and *O. clelandii* "Iowa, 1974, *Allbaugh s.n.*" (ring of 14 chromosomes) as staminate parent (DUSS cult. no. 82–711) resulted in two morphologically almost identical phenotypes. The first type had spreading calyx pubescence, relatively broad leaves, and a densely flowered spike, whereas the plants of the second type had appressed calyx pubescence, small narrow leaves, and a fewer-flowered, more open spike. However, the differences between these phenotypes are minimal and suggest that the two complexes were derived from one species rather than from two different species following hybridization. Also, the differences of the two F_1 phenotypes are within the variation range of *O. rhombipetala* itself.

Crosses between *O. clelandii* and *O. rhombipetala* also showed that *O. clelandii* transfers both chromosomal complexes through the ovule and the pollen. If *O.*

clelandii is used as the pistillate parent, both complexes occur in a proportion of approximately 1:1, but when used as the staminate plant, the broad-leaved type tends to dominate (DUSS cult. no. 79–171, 79–172, 79–175, 79–176).

5. Oenothera curtissii Small, Fl. S.E. U.S. ed. 2, 1353, 1375. April 1913. *Oenothera heterophylla* var. *curtissii* (Small) Fosberg, Amer. Midl. Naturalist 27: 763. 1942.—TYPE: U.S.A. Georgia: Decatur Co., in dry open places in valley of Flint River at Bainbridge, 22 Aug 1901, *Curtiss 6880* (holotype: US-394775!; isotypes: CU! E! G! GH! HBG! K! KSC! MIN! MO! NEB! NY! P! PRC! UC! US!).—Small intended his binomial to be a new combination based on Rose's name (*Raimannia curtissii* Rose, Contr. U.S. Natl. Herb. 8: 330. 1905); however, the genus *Raimannia* was not validly published by Rose in 1905, and therefore the species proposed by Rose were also not validly published (ICBN, Art. 43.1). Since Small's protologue only gives reference to Rose's publication, the US sheet of *Curtiss 6880* may be considered the holotype.

Erect annual, biennial, or perhaps short-lived perennial herbs, forming a rosette; stems 3–8 dm long (longer in cultivation), usually green, densely to sparsely strigillose, simple or branched. Rosette leaves 7–17 cm long, 0.5–1.5 cm wide, narrowly oblanceolate, deeply lobed with obtuse lobes to subentire, apex acute, gradually narrowed to the petiole; cauline leaves 2–8 cm long, 0.5–1.5 cm wide, very narrowly elliptic or narrowly oblong to oblong, bluntly dentate to subentire, apex acute, base obtuse to narrowly cuneate, sessile or short-petiolate; bracts 0.6–1.7 cm long, 0.3–0.5 cm wide, narrowly ovate to ovate or oblong to broadly oblong, apex acute, base truncate to rounded, subentire, sessile; all leaves and bracts green, strigillose. Inflorescence lax, usually without lateral branches. One to two flowers per spike opening each day near sunset. Floral tube 2.3–3.7 cm long, ca. 1 mm in diameter, sometimes flecked with red, strigillose, sometimes also sparsely glandular-puberulent. Mature buds 2–3 mm in diameter, narrowly oblong to oblong in outline, not overtopping apex of the spike. Sepals 0.7–1.3 cm long, yellowish, sometimes flecked with red, densely to sparsely strigillose, sometimes also sparsely glandular-puberulent, sepal tips 0.3–0.8 mm long, erect or somewhat spreading, strigillose. Petals 0.8–1.7 cm long, 0.6–1.2 cm wide, yellow, broadly elliptic to suborbicular, apex acute to almost rounded. Filaments 6–10 mm long; anthers 1.5–4 mm long, shedding pollen directly on the stigma at anthesis, pollen ca. 50% fertile. Ovary 0.8–1 cm long, ca. 1 mm in diameter, densely strigillose. Style 3–4.5 cm long, the exserted part 8–11 mm long; stigma surrounded by the anthers at anthesis, lobes 1.5–2 mm long. Capsule 1–2.5 cm long, 2–3 mm in diameter, narrowly lanceoloid, strigillose. Seeds 1–1.3 mm long, 0.5–0.7 mm in diameter, narrowly to broadly ellipsoid, brown, often dark flecked, the surface pitted. Autogamous, permanent structural heterozygote. Chromosome number: n = 7 (⊙ 14* chromosomes at meiotic metaphase I). Fig. 3.

Phenology. Flowering from June to September.

Distribution (Fig. 2). Dry sandy places in woods, fields, and along roadsides in southwestern Alabama, northern Florida, southern Georgia, and southern South Carolina.

SPECIMENS FROM CULTIVATED PLANTS. U.S.A. FLORIDA: Wakulla Co., W of Wakulla, jct. U.S. Rte 319 & Fla Rte 267, *Godfrey 76646* [cult. no. 80–302*] (DUSS, M, MO).—GEORGIA: Baker Co., Flint River, 13 mi SSW of Newton, *Jones & Jones 22536* [cult. no. 76–711*] (DUSS, M, MO).

ADDITIONAL SPECIMENS EXAMINED. U.S.A. ALABAMA: Baldwin Co., 5 mi SE of Loxley, *Corbin 168* (SMU).—FLORIDA: Duval Co., Girwin road, *Creager & Beckner 655* (FLAS); vicinity of Mayport and Jacksonville, *Keeler 1870–1876* (NY). Leon Co., along forest road 301, ca. 1 mi W of Fla Rte 260, *Henderson 64-271* (DS, FSU, KANU, VT); Tallahassee, along airport road, *Kurz 1601* (FLAS). Madison Co., *Hitchcock 671* (F). Marion Co., McIntosh, *Meebold 27700* (M). Okaloosa Co., ca. 12 mi S of jct. Hwy 255 along Hwy 251, ca. 12 mi W of Fort Walton Beach (T2S, R25W, Sec. 5), *Smith 1569* (FLAS, UNCC); (T1S, R23W, Sec. 20), *Tyson 75* (FLAS); (T1S, R23W, Sec. 34), *Tyson 63* (FLAS). Putnam Co., Welaka, *Laessler 171* (FLAS). Santa Rosa Co., along Hwy 98 between Pensacola and Fort Walton, *Farlin 12480* (NA). Suwannee Co., 5 mi N of Live Oak, *Arnold & West 171* (FLAS); 1 mi NW of Live Oak, *Wiegand & Manning 2208* (BH, GH); E of Ellaville, *Ramsey et al. 277* (FSU). Wakulla Co., 2 mi S of Crawfordville, *Dress & Read 7747* (BH). Walton Co., Portland, *Godfrey 64265* (DS, FSU). County unknown: East Pass, *Tracy 6614* (E, G, NY, US).—GEORGIA: Baker Co., along Flint River, SSW of Newton below mouth of Ichawaynochaway Creek, *Jones & Jones 22536* (GA, MO), *Thorne 5744* (BH, GA, RSA), *Duncan 6657* (GA, NCU); Flint River near Bainbridge, 1884, *Chapman s.n.* (AUA, F, MIN, MO, NOSU, US); Emory University Field Station, *Thorne & Davidson 16410* (IA). Lanier Co., 3 mi N of Lakeland, *McKellar 28* (GA).—SOUTH CAROLINA: Allendale Co., 0.8 mi NE jct. Co. Rte 60 & Rte 25 on Rte 25, W of Barton, *Ahles & Bell 15877* (FSU, UNCC, USCH).

Like *O. clelandii, O. curtissii* is a small-flowered, autogamous, permanent structural heterozygote forming a ring of 14 chromosomes at meiotic metaphase I. Geographically it is completely isolated from both *O. rhombipetala* and *O. clelandii.* It differs primarily in its considerably looser, few-flowered inflorescence, with no more than two flowers open per day on any one spike; in *O. clelandii* and *O. rhombipetala,* up to five or more flowers may open per day. Also, unlike *O. clelandii, O. curtissii* is a short-day plant, a feature that causes considerable difficulties in cultivation at higher latitudes than those at which it normally occurs. In order for it to flower in cultivation the first year, it must be grown in a short-day situation and must be coddled carefully through the winter; in the second year, it often does not bloom until September.

Oenothera curtissii presumably evolved from *O. rhombipetala* independently from *O. clelandii,* another structural heterozygote. In artificial crosses with *O. heterophylla,* it behaved similarly to *O. clelandii:* apparently both complexes are transmitted through the ovule and the pollen (DUSS cult. no. 79–179, 79–184, 79–185, 79–186). Therefore, *O. clelandii* and *O. curtissii* appear to be isogamous and independently derived from morphologically similar but chromosomally differentiated individuals within *O. rhombipetala.*

IB. Oenothera section **Oenothera** subsection **Raimannia** series **Raimannia** (Rose ex Britton & A. Brown) W. Dietrich & W. L. Wagner, Ann. Missouri Bot. Gard. 74: 148. 1987.

Erect to procumbent annual or perennial herbs forming rosettes; stems green or flushed with red, simple or much-branched, strigillose, or strigillose and villous, especially in the region of the inflorescence sometimes also glandular-puberulent. Rosette leaves very narrowly oblanceolate to oblanceolate or lanceolate, parted to dentate, acute; cauline leaves narrowly oblanceolate to oblanceolate, narrowly elliptic, elliptic, narrowly obovate or narrowly oblong, parted to subentire, lobes more or less dentate, apex acute, base narrowly cuneate to acute to almost sessile;

all leaves strigillose, or strigillose and villous, sometimes also glandular-puberulent. Inflorescence lax, often with lateral branches. Mature buds with floral tube curved upward. Usually only one flower per spike opening each day near sunset. Petals yellow, sometimes pale yellow fading red to orange when wilted, very broadly obovate, truncate to emarginate at apex. Stigma elevated above the anthers at anthesis or surrounded by the anthers and pollen shed directly on it. Capsule cylindrical. Seeds ellipsoid to subglobose, brown, sometimes with darker flecks (*O. drummondii, O. humifusa*).

Series *Raimannia* of subsect. *Raimannia* consists of six species that are often highly variable; however, only *O. drummondii* can be subdivided into two subspecies that are separated geographically. Typical of the species of ser. *Raimannia* are the loose inflorescences, which often have lateral branches, and the upward-curving flower buds. Further comparisons with ser. *Candela* were made in the discussion of that series.

The distribution of ser. *Raimannia* is essentially the same as that of ser. *Candela,* but it extends farther east, to the Atlantic coast, and south to the state of Campeche, Mexico, along the Gulf of Mexico. *Oenothera drummondii* subsp. *thalassaphila* is disjunct, occurring along the Pacific coast at the southern tip of Baja California.

Oenothera grandis, O. falfurriae, O. mexicana, and *O. drummondii* are bivalent-forming species, whereas *O. laciniata* and *O. humifusa* are permanent structural heterozygotes. In this series, only *O. grandis* is self-incompatible; all other species, both bivalent formers and complex structural heterozygotes, are self-compatible and largely autogamous.

6. Oenothera grandis (Britton) Smyth, Trans. Kansas Acad. Sci. 6: 160. 1899. *Oenothera sinuata* var. *grandiflora* S. Watson, Proc. Amer. Acad. Arts 8: 581, 614. 1873. *Oenothera sinuata* var. *grandis* Britton, Mem. Torrey Bot. Club 5: 358. 1894. *Oenothera laciniata* var. *occidentalis* Small, Bull. Torrey Bot. Club 23: 173. 1896, nom. superfl. *Oenothera laciniata* var. *grandis* (Britton) Britton in Britton & A. Brown, Ill. fl. n. U.S. 2: 487. 1897. *Raimannia grandis* (Britton) Rose ex Britton & A. Brown, Ill. fl. n. U.S., ed. 2, 2: 597. 1913. *Oenothera laciniata* var. *grandiflora* (S. Watson) Robinson, Rhodora 10: 34. 1908.—Type: U.S.A. Texas: [Tarrant Co.], sandhills near Austin, 13 May 1872, *Hall 203* (lectotype, here designated: GH!; isolectotypes: BM! MO! NA! NY! POM! US!).—Among the three collections cited by Watson (*Hall 203; Palmer 132; Wright s.n.*), Munz (Amer. J. Bot. 22: 656. 1935.) suggested taking *Palmer, E. 132* (GH, US) from False Washita, Caddo Co., Oklahoma, as the lectotype. But since Watson's protologue mentions only Texas (p. 581), and Munz was not definite, only saying ". . . if we take the first of three specimens cited by Watson, Namely *Palmer 132*", it seems best to designate a lectotype here.

Erect to ascending annual herbs, forming rosettes; stems 1.5–6 (–10) dm long, usually green, simple or with arcuating lateral branches arising from the rosette, strigillose and sparsely villous, becoming glandular-puberulent in the upper part. Rosette leaves 5–13 cm long, 1–3 cm wide, narrowly oblanceolate, lobed to dentate, gradually narrowed to the petiole; cauline leaves 3–10 cm long, 1.5–3.5 cm wide, narrowly obovate or oblanceolate to narrowly elliptic, lobed or dentate, the lobes often dentate, narrowly cuneate at base, short-petiolate to almost sessile; bracts 2–9 cm long, 0.8–4.5 cm wide, narrowly lanceolate to lanceolate or narrowly ovate to elliptic, lobed to dentate, rounded to narrowly cuneate at base, sessile; leaves and bracts green, strigillose and villous. Inflorescence lax, often with lateral branches. Usually one flower per spike opening each day near sunset. Floral tube 2.5–4.5 cm long, 1.5–2 mm in diameter, curved upward, densely to sparsely villous and glandular-puberulent, sometimes sparsely strigillose in lower part. Mature buds 5–10 mm in diameter, lanceoloid. Sepals 1.5–3 cm long, green to yellowish, often red-striped at the junction of the sepals with the floral tube, sometimes red-flecked, villous and glandular-puberulent, sepal tips 1.5–5 mm long, erect or hornlike in bud, strigillose and villous. Petals 2.5–4 cm long, 3–5.5 cm wide, yellow, very broadly obovate, apex truncate to slightly retuse. Filaments 12–22 mm long; anthers 4–11 mm long, pollen ca. 90–100% fertile. Ovary 1.2–2.5 cm long, 1.5–1.8 mm in diameter, strigillose and villous, sometimes also glandular-puberulent in upper part. Style 4–7.5 cm long, the exserted part 1.5–3 cm long; stigma elevated above the anthers at anthesis, lobes 5–13 mm long. Capsule 2.5–5 cm long, 2–3 mm in diameter, cylindrical, strigillose and villous. Seeds 0.8–1.5 mm long, 0.5–0.9 mm in diameter, broadly ellipsoid to subglobose, brown, immaculate, the surface pitted. Self-incompatible. Chromosome number: $n = 7$ (7_{II}*, or ⊙4 and 5_{II}** at meiotic metaphase I). Fig. 4.

Phenology. Flowering from March to September.

Distribution (Fig. 5). Open sandy sites from Texas, eastern New Mexico, Oklahoma, Kansas, Colorado, Nebraska, Louisiana, Arkansas, and Missouri; also northeastern Tamaulipas, Mexico. Localities in California, Florida, Alabama, Illinois, North Carolina, Maryland, New Jersey, and Connecticut probably represent introductions. This species has also been reported to be naturalized in Japan (Asai 1973).

Specimens from Cultivated Plants. U.S.A. Kansas: Clark Co., near Ashland, *Brooks 6517* [cult. no. 75–1473*] (DUSS, M, MO). Comanche Co., 4 mi E of Protection, *Brooks 6564* [cult. no. 75–1474*] (DUSS, M, MO). Edwards Co., 4 mi N of Fellsburg, *Brooks 6315* [cult. no. 75–1480d*] (DUSS, M, MO); 4.5 mi N of Fellsburg, *Brooks 6316* [cult. no. 75–1476*] (DUSS, M, MO); 5 mi E of Lewis, *Brooks 6325* [cult. no. 77–566**] (DUSS, M, MO); 2 mi W of Trousdale, *Brooks 6313* [cult. no. 75–1480*] (DUSS, M, MO). Kiowa Co., 2 mi E of Brenham, *Brooks 6616* [cult. no. 75–1478*] (DUSS, M, MO). Pawnee Co., 6 mi S of Larned, *Brooks 6311* [cult. no. 75–1480b*] (DUSS, M, MO).—Texas: Kenedy Co., Sarita, 1965, *Cleland 452* [cult. no. 77–562*] (DUSS, M, MO).

Representative Specimens. Mexico. Tamaulipas: Punta Piedras, 50 mi SE of San Fernando, *Fryxell 1230* (CAS, ENCB); San José, *LeSueur 319* (F), *LeSueur 321* (F, GH, TEX).——U.S.A. Alabama: Colbert Co., Leighton, 1912, *Howell s.n.* (US).—Arkansas: Faulkner Co., *Demaree 5637* (NY). Logan Co., 1925, *Young s.n.* (PH). Pulaski Co., Little Rock, 1886, *Hasse s.n.* (MU). Yell Co., Greenville, 1924, *Young s.n.* (CM).—California: San Bernardino Co., 1906, *Robertson s.n.* (DS).—Colorado: El Paso Co., Colorado Springs, *Biltmore 1383* (US). Sedgwick Co., 1 mi S of Julesburg, *Stephens & Brooks 24058* (UWL).—Connecticut: Hartford Co., 1919, *Hanmer s.n.* (CONN).—Florida: Polk Co., Bartow, 1931, *McFarlin s.n.* (CAS, MICH).—Illinois: Champaign Co., Urbana, 1945, *Boewe s.n.* (ILL.). Cook Co., Chicago, 1893, *Williamson s.n.* (PENN).—Indiana: Porter Co., Kelsor, 1923, *Lyon s.n.* (MICH).—Kansas: Barber Co., Sun City, *Wagenknecht 2456* (KANU). Barton Co., Great Bend, 1947, *Wilson s.n.* (KSC). Cheyenne Co., St. Francis, *Stellar 9* (KSC). Clark Co., Ashland, *Brooks 6517* (MO). Comanche Co., Coldwater, *McGregor 12648* (KANU). Edwards Co., Kinsley, 1937, *Bondy s.n.* (ARIZ, CAN, CAS, CM, DUKE, F, GH, MIN, NMC, OKL, TENN, UARK, UMO, US, VDB, WIS, WVA). Ellsworth Co., Wilson, *Weber 32* (KSC). Finney Co., Garden

FIG. 4. *Oenothera grandis* and *O. falfurriae*. *O. grandis:* a. Habit (*Brooks 5863*). b. Bud, bracts, and capsule (*Benbow 81*). *O. falfurriae* (*Shaw & Allred 2020,* cult. no. 81–116, 81–117; MO): c. Inflorescence. d. Bud. e. Sepal tips in bud.

City, *Stephens 11281* (DS, KANU, NCU). Ford Co., Dodge City, *Stephens 47785* (KANU). Grant Co., Ulysses, *McGregor 25512* (KANU). Harper Co., Harper, *Gates 15172* (KSC). Harvey Co., Hesston, *Brooks 5812* (KANU). Haskell Co., Sublette, *McGregor 25535* (KANU). Hodgeman Co., Hanston, *Brooks 6322* (KANU). Kingman Co., Kingman, *Stephens 53965* (KANU). Kiowa Co., Greensburg, 1888, *Kellerman s.n.* (KSC). Pawnee Co., Larned, *McGregor 16914* (KANU, NY, SMU). Pottawatomie

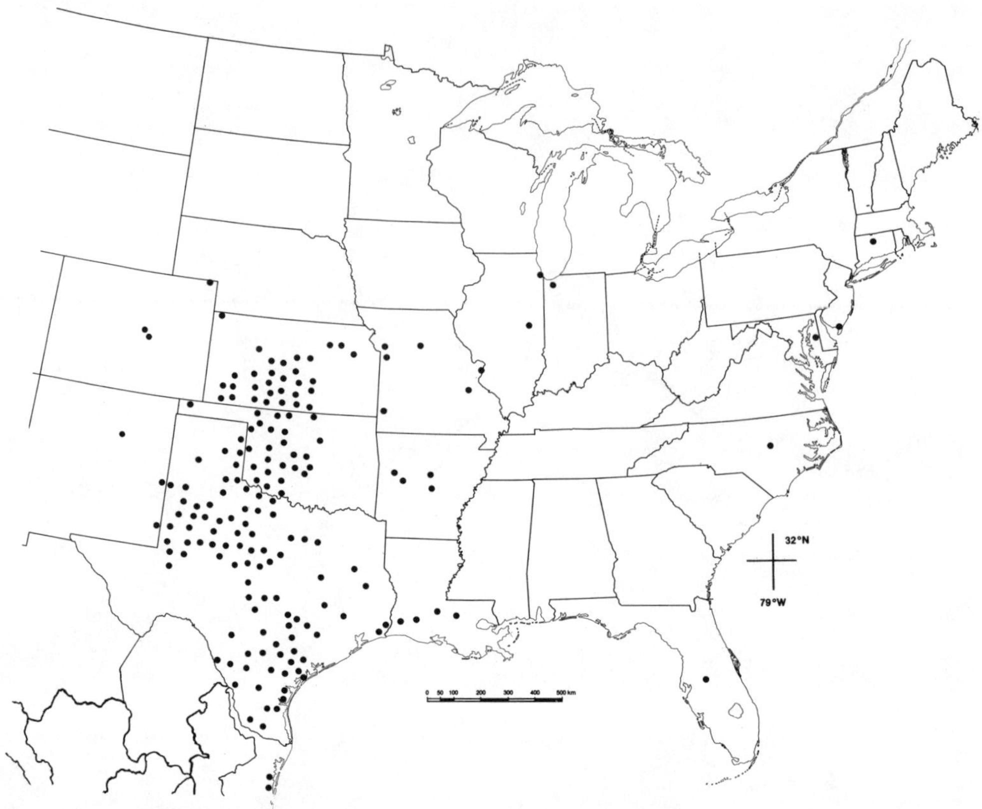

FIG. 5. Distribution of *Oenothera grandis.*

Co., St. George, 1894, *Norton s.n.* (KSC). Pratt Co., Cairo, *Stephens 47638* (KANU, OKL). Reno Co., Sylvia, *Brooks 5863* (MO); Hutchinson, *Benke 4287* (F, US). Rice Co., Raymond, *Stephens 65232* (KANU). Riley Co., 1895, *Asbury s.n.* (?). Rush Co., Bison, *McGregor 27260* (KANU). Saline Co., *Hancin 1270* (KANU). Sedgewick Co., Goddard, *Brooks 6779* (KANU). Seward Co., Liberal, *Stephens 11212* (DS, KANU). Shawnee Co., Topeka, 1903, *Schaffner s.n.* (OS). Stafford Co., S 281–50 jct. on Hwy 50, *Harms 1090* (KANU, NY, UC). Stevens Co., Moscow, *McGregor 25467* (KANU). Sumner Co., Belle Plain, *Brooks 6762* (KANU). Trego Co., Cedar Blugg Reservoir, *Brooks 6012* (KANU).—LOUISIANA: Calcasieu Pa., Lake Charles, *Daves s.n.* (F). East Baton Rouge Pa., Baton Rouge, 1934, *Brown s.n.* (LSU). Jeff Davis Pa., Welsh, *Palmer 7647* (CAS, K, MO, NY, US). St. Landry Pa., Lawtell, *Brown 6254* (LSU).—MARYLAND: Caroline Co., Greensborough, *Shreve 1640* (US).—MISSOURI: Chariton Co., 1920, *Young s.n.* (CM). Clay Co., *Mackenzie 124* (MIN). Jackson Co., Sheffield, *Bush 551* (MIN, NA, US, VT). Jasper Co., Joplin, *Palmer 550* (MO). St. Louis Co., St. Louis between Victor & Rutger Sts, *Mühlenbach 63* (MO). Washington Co., Irondale, 1898, *Russell s.n.* (MO).—NEW JERSEY: Cape May Co., Cold Spring, 1923, *Brown s.n.* (PH).—NEW MEXICO: Lea Co., Hobbs, *Pearce 2640* (SMU). Roosevelt Co., Portales, *Weber 14448* (COLO, NCU). San Miguel Co., El Pueblo Natl. Forest, 14 mi S of Las Vegas, 1975, *Schooley s.n.* (COLO).—NORTH CAROLINA: Johnston Co., Princeton, 1936, *Mitchell s.n.* (DUKE).—OKLAHOMA: Alfalfa Co., Salt Plains Lake, *Kelting 238* (OKL). Blaine Co., Geary, *Engleman 1397* (OKL). Caddo Co., Ft. Cobb, *Goodman 6255* (GH, KANU, OKL, RSA, UC). Canadian Co., Hinton, *Hopkins 1397* (OKL). Cimarron Co., Boise City, *Jeffs 55* (OKL). Cleveland Co., Norman, *Little 104* (OKL). Comanche Co., Medicine Park, 1925, *Myers s.n.* (OKL). Cotton Co. Devol, *Stephens 20401* (KANU, NY). Custer Co., Weatherford, 1930, *Jeffs s.n.* (OKL). Ellis Co., Shattuck, *Clifton 3041* (DS, NY). Grady Co., 1957, *Buck s.n.* (OKL). Greer Co., Mangum, *Bull 113* (OKL). Harmon Co., Hollis, *Waterfall 16934* (OKL). Harper Co., Buffalo, *Goodman &*

Lawson 8306 (G, KANU, NCU, OKL). Jackson Co., Altus, *Waterfall 17461* (KANU, NCU, OKL, USF). Kay Co., Tonkawa, *Johnson 90* (F). Kiowa Co., near Kiowa-Caddo Co. line on Hwy 9, *Lawson & Massey 293* (NCU, OKL, SMU, TEX). Major Co., Cleo Springs, *Waterfall & Coryell 9918* (ARIZ, KANU, MASS, OKLA, RSA, SMU, TAES, TEX). McClain Co., W of Canadian River Bridge, *Nelson 39* (OKL). Oklahoma Co., Oklahoma City, *Waterfall 427* (OKLA). Payne Co., Ripley, *Raffaelli 35* (ARIZ, KANU, SMU). Roger Mills Co., Durham, *Taylor & Baalman 2618* (OKL). Swanson Co., Snyder, *Stevens 1194* (P), *Stevens 1195* (GH, MIN, NY, OKLA). Tillman Co., Davidson, *Shinners 2606* (SMU, WIS). Washita Co., *Eskew 4619* (BH, OKL). Woods Co., Alva, *Stevens 3021* (DS, ILL, MIN, NY, OKL). Woodward Co., Sharon, *Stephens 74738* (KANU, OKL).—TEXAS: Andrews Co., Andrews, 1958, *Scudday s.n.* (TEX). Angelina Co., Lufkin, 1951, *Price s.n.* (ASTC). Aransas Co., Rockport, *Whitehouse 18255* (MICH, NY, SMU). Archer Co., between Megargel & Olney, *Whitehouse 9646* (MICH, NY, SMU). Atascosa Co., Pleasanton, *McCullough 59* (NY, OKL). Bailey Co., Muleshoe, *Rosson 1740* (TTC). Bastrop Co., Bastrop, *Cory 51673* (DS, MICH, SMU, US). Baylor Co., Seymour, *Shinners 18595* (SMU). Bexar Co., Thelma, *Johnson & Webster 573* (GH, NY, OKLA, PH, TEX, US). Brazos Co., Brushy Creek, *Reeves 2009* (SMU, TAES). Brooks Co., 14 mi W of Falfurrias, *Watson 60* (SMU). Brown Co., Brownwood, *Latham 15* (OKLA, TAES, TEX). Burnet Co., Fairland, 1898, *Bray s.n.* (TEX). Caldwell Co., Fentress, 1926, *Hill s.n.* (TEX). Callahan Co., Clyde, *Henderson 63–572* (FSU, SMS, SMU, TEX, VT). Cherokee Co., Gallatin, 1958, *Garner s.n.* (ASTC). Childress Co., Childress, *Correll & Johnston 16854* (TEX). Coke Co., Silver, *Rowell 16011* (CAN, MO). Collingsworth Co., Wellington, *Bailey & Bailey 10149* (BH). Comanche Co., Comanche, 1900, *Eggert s.n.* (MIN, MO). Crane Co., Crane, *Warnock 14633* (TEX). Crosby Co., Crosbyton, *Smith 17* (DS). Dallas Co., Dallas, *Reverchon 297* (P), *Reverchon 905a* (CU, F, GH, KANU, MASS, MIN, NY, PH, US, VT, YU). Dawson Co., Lamesa, *Shinners 31809* (SMU, VDB). DeWitt Co., Cuero, *Blanch 72* (TEX). Dimmit Co., Catarina, *de Anda 50* (FSU, GA, SMU, TEX). Duval Co., 9 mi W of Falfurrias, *Ramirez et al. 8783* (OKLA, SMU, TAES, TEX). Eastland Co., Cisco, *Shinners 26407* (SMU). Ector Co., 22 mi W of Midland, *Flyr 1430* (MO, SMU). Erath Co., *Gough 224* (TEX). Fayette Co., La Grange, 1894, *Schneck s.n.* (ILL). Frio Co., Pearsall, *Gregory 284* (NCU, UC, US). Gaines Co., Seminole, *Bryan et al. 15* (TTC). Garza Co., Post, *Hutchins 1992* (SMU, TEX, TTC, UNM). Gillespie Co., Fredericksberg, *Palmer 10054* (DS, MIN, US). Goliad Co., Goliad, *Williams 109* (F, PH, TEX). Gonzales Co., Palmetto State Park, *Rose-Innes 41005* (TEX). Gray Co., 17 mi SE of Lefors, *Stephens 76475* (KANU, VDB). Hall Co., Turkey, *Higgins 7181* (NY). Hardeman Co., Chillicothe, *Cory 13362* (POM). Haskell Co., Rull, *Cory 37211* (TAES). Hemphill Co., Canadian, *Rowell 5804* (SMU, TAES, TEX). Hidalgo Co., San Manuel, *Runyon 2569* (CAS, POM, TEX, WIS). Hockley Co., Whitharral, *Stephens 72996* (KANU). Howard Co., Big Spring, *Gregory 421* (CAS, NCU, RSA, UC). Jeff Davis Co., Ft. Davis, 1902, *Bray s.n.* (TEX). Jefferson Co., China, *Hunt 3* (DS). Jones Co., Anson, *Henderson 63–147* (FSU, SMS, SMU, TAES, TEX, VT). Karnes Co., E of San Antonio RR crossing on Hwy 72, *Johnson 1110* (RSA, TAES). Kenedy Co., Sarita, *Lundell & Lundell 10849* (MICH, POM, SMU). Kent Co., Jayton, *Walker 16* (TTC). Kleburg Co., Riviera, *Gorgoia et al. 8815* (OKLA, TEX). Knox Co., jct. of Hwys 283 & 1292, *McGrary & McGrary 37* (TEX). Lamb Co., Amherst, *Stephens 80326* (KANU). Lasalle Co., Cotulla, *Wiegand & Wiegand 1572* (PH). Limestone Co., Tehuacana, *McBryde 3034* (NY, SMU). Live Oak Co., 20 mi SW of Georgewest, *Cantu 61* (DUKE, FSU, GA, SMU). Llano Co., Enchanted, *Innes & Warnock 787* (F, GH, TEX). Lubbock Co., Lubbock, *Benbow 81* (MO); *Demaree 7702* (DS, GH, MO, SMU, TENN, TEX, TTC, UARK, US, WIS). Lynn Co., 25 mi SE of Tahoka, *Bauer 103* (OKLA). Martin Co., Tarzan, *Shinners 33040* (SMU). Mason Co., Mason, *Shinners 26300* (SMU). Maverick Co., 20 mi SE of Eagle Pass, *Cuellar 48* (DUKE, SMU, TEX, TTC). McCulloch Co., Brady, *Munz & Gregory 23432* (CAS, RSA, UC). Medina Co., Devine, *de Los Santos 39* (DUKE, FSU, GA, SMU). Midland Co., Chub, *Raven & Gregory 19215* (DAO, DS, RSA, US). Mitchell Co., Lavaca Nav. Co., *Pohl 4906* (PENN, SMU). Montgomery Co., Willis, *Warner s.n.* (MO). Motley Co., Matador, 1934, *Tharp s.n.* (TEX). Nolan Co., Sweetwater, *Demaree 40871* (NCU). Nueces Co., Mustang Island, *Gillespie 207* (TEX). Orange Co., Orange, *Wild Wood Club 28* (TEX). Parker Co., Millsap, 1927, *Wadsworth s.n.* (TEX). Pratt Co., Preston, *McGregor 12482* (KANU). Randall Co., 11 mi SE of Randall, *Stewart 47* (TTC). Refugio Co., Greta, *Lundell & Lundell 10855* (MICH, POM, SMU). Runnels Co., halfway between San Argelo & Abilene, *Clayton 2330* (MIN). San Augustine Co., Calgary, *Reed 3384* (US). San Patricio Co., Welder Wildlife Refuge, *Williges 6–C* (NCY, TEX). San Saba Co., 1931, *Norton s.n.* (TEX). Scurry Co., 10 mi E of Sweetwater, *Hooper 2* (FLAS). Shackleford Co., Leuders, *Henderson 62–235* (FSU). Starr Co., Falcon State Park, *Strother 90* (TEX). Stonewall Co., 10 mi NW of Aspermat, *Dieckmeter 11* (SMU, TAES, TEX). Tarrant Co., Fort Worth, *Hunnewell*

5929 (GH). Taylor Co., Camp Barkeley, *Tolstead 7025* (ISC, MICH, MO, PENN, SMU, TEX, US). Terry Co., Brownfield, *Wolff 1987* (F, US). Travis Co., Austin, 1909, *York s.n.* (TEX). Uvalde Co., 130 mi NW of Laredo, *Guerra et al. 46* (OKL, TEX). Victoria Co., Victoria 1900, *Eggert s.n.* (MIN, MO). Waller Co., Hempstead, 1872, *Hall s.n.* (F). Webb Co., Drake Ranch, *Drake 29* (OKLA). Wheeler Co., Kellerville, *Stephens 76552* (KANU). Wichita Co., Electra, *Whitehouse 9782a* (MICH, SMU). Wilbarger Co., Odell, *Stephens 20694* (DS, KANU). Wilson Co., Stockdale, *Cory 54072* (GA, KANU, NSC, SMU, TEX).

SPECIMEN FROM OUTSIDE NATURAL AREA. GERMANY. Hamburg, Reiherstieg, 1909, *Schmidt s.n.* (HBG).—NETHERLANDS. Wormerveer, 1923, *Kloos s.n.* (L); Deventer, 1911, *Brand s.n.* (L).

The geographical ranges of *O. grandis* and *O. rhombipetala* are strikingly similar, which suggests they have similar ecological requirements. *Oenothera grandis* and *O. drummondii* have the largest flowers in ser. *Raimannia,* reaching a diameter up to 9 cm. *Oenothera grandis* differs from *O. drummondii* by its deeply incised leaves, the lobes usually serrated, and sparse pubescence of erect hairs. In contrast, *O. drummondii,* as a rule, has dense, appressed pubescence. The size of the flowers and the fact that *O. grandis* is a self-incompatible, bivalent former (7_{II} in meiotic metaphase I) separate it very clearly from *O. laciniata,* which very often is similar in its habit but has smaller flowers and is an autogamous, permanent structural heterozygote. Because the breeding systems of these plants were unknown, *O. grandis* had most often been judged to be a large-flowered variety of *O. laciniata.* At the present time, the only chromosome arrangements known in *O. grandis* are 7_{II}, but one collection (*Brooks 6325* from Edwards County, Texas) had 5_{II} + ⊙4. Further sampling would probably reveal other small translocations.

Pollination has been observed once in *O. grandis* (Big Spring, Howard Co., Texas) by Gregory (1963, 1964). *Hyles lineata* (Fabricius) was the principal pollinator at the Big Spring site, but *Manduca quinquemaculata* (Haworth) and the nocturnal bee, *Lasioglossum (Sphecodagastra) texanum* (Cresson) were collected at a light trap and probably also pollinate *O. grandis,* the bee probably being responsible for relatively few pollinations. Presumably *O. drummondii* subsp. *drummondii,* which also grew at this site, shared the same pollinators.

7. Oenothera falfurriae W. Dietrich & W. L. Wagner, Ann. Missouri Bot. Gard. 74: 149. 1987.—TYPE: Grown from seeds and cultivated in the Botanical Garden of Düsseldorf, Germany, 2 Jul 1981, cult. no. 81–115 from seeds collected in U.S.A., Texas, Brooks Co., 13.3 mi S of junction of Highways 281 and 285 in Falfurrias, 10 May 1978, *Allred & Shaw 2021* (holotype: MO-3332203!; isotypes: DUSS! M! MO!).

Erect to decumbent annual herbs, forming rosettes with only a few leaves; stems 1–4 dm long, usually simple, moderately to sparsely strigillose, villous, sometimes also glandular-puberulent. Rosette leaves 5–12 cm long, 1.3–3.5 cm wide, oblanceolate, deeply lobed to dentate or sometimes some of them subentire, apex acute, gradually narrowed to a short petiole; cauline leaves 2–8.5 cm long, 1–3 cm wide, narrowly oblanceolate to elliptic or narrowly lanceolate, usually dentate, occasionally pinnatifid or subentire, apex acute, gradually narrowed to subsessile base; bracts 2–4.5 cm long, 0.5–2.5 cm wide, spreading, elliptic, narrowly ovate to lanceolate, flat, dentate or subentire to pinnately lobed, narrowed to the base, subsessile; leaves and bracts moderately to sparsely villous and glandular-puberulent primarily on the midrib of the lower surface and along the margin, usually also sparsely to moderately strigillose. Inflorescence lax, simple or with lateral branches. Usually only one flower per spike opening each day near sunset, erect at anthesis. Floral tube 2.5–4 cm long, densely to sparsely villous and glandular-puberulent. Mature buds 0.4–0.6 cm in diameter at the base, narrowly oblong to oblong, rarely lanceoloid. Sepals 1–2.2 cm long, green to greenish yellow, sometimes with red spots, pubescence same as the floral tube, sepal tips 0.5–2 mm long, erect in bud, strigillose and villous. Petals 1.3–2.5 cm long, 1.4–2.7 cm wide, yellow, broadly obovate, apex truncate to slightly retuse. Filaments 10–17 mm long; anthers 4–5 mm long, pollen ca. 90–100% fertile. Ovary 1–1.7 cm long, ca. 1.5 mm in diameter, densely villous, strigillose, and glandular-puberulent. Style 3.5–5 cm long, the exserted part 1.2–2 (–2.5) cm long; stigma usually slightly elevated above the anthers at anthesis, lobes 3–7 mm long. Capsule cylindrical, 2–4.5 cm long, 2–2.5 mm in diameter. Seeds 0.8–1.4 mm long, 0.3–0.6 mm in diameter, ellipsoid, brown, the surface pitted. Self-compatible, modally autogamous. Chromosome number: n = 7 (7_{II}* at meiotic metaphase I). Fig. 4.

Phenology. Flowering from April through August.

Distribution (Fig. 6). Open sandy sites in southeastern Texas.

Specimens from Cultivated Plants. U.S.A. Texas: Brooks Co., 11.7 mi S jct. Hwys 281 & 285, Falfurrias, *Shaw & Allred 2020,* pro parte [cult. no. 80–311*, 80–314*, 80–316*, 81–116, 81–117] (DUSS, M, MO); 13.3 mi S jct. Hwys 281 & 285, Falfurrias, *Shaw & Allred 2021,* pro parte [cult. no. 79–138*, 80–326*] (DUSS, M, MO). Kenedy Co., Sarita, *Shaw & Allred 2016,* pro parte [cult. no. 79–129*, 80–309*, 80–341*] (DUSS, M, MO).

Additional Specimens Examined. U.S.A. Texas: Aransas Co., Rockport, *Chase 6043* (MICH). Brazos Co., College Station, 1891, *Jennings s.n.* (MICH). Brooks Co., 11.7 mi S of jct. Hwys 281 & 285 in Falfurrias, *Shaw & Allred 2020* (MO); 13.3 mi S of jct. Hwys 281 & 285, Falfurrias, *Shaw & Allred 2021* (MO); San Manuel below Falfurrias, 1930, *Tharp s.n.* (TEX); N of Encino, Hwy 281, *Runyon 2646* (POM, TEX). Cameron Co., La Feria, 1926, *Cannon s.n.* (MICH); Resaka de la Palma, *Runyon 539* (GH, US); Resaka Park near Brownsville, *Runyon 2486* (CAS, POM, TEX, WIS), *Runyon 6042* (TEX), *Runyon 4922, 4293* (WIS). Frio Co., Rio Frio N of Dilley, *Lucas 14203* (LL). Harris Co., Houston, *Traverse 388* (SMU). Hidalgo Co., 15 mi N of Edinburg, Hwy 66, *Runyon 3111* (TEX); La Joya, 1942, *Walker s.n.* (MO, TEX, 84% fertility). Jim Hogg Co., 27 mi S of Hebronville, farm road 1027, *Pena 8* (SMU). Jim Wells Co., Hwy. 44, 1 mi W of Alice, *Mahler 5244* (SMU). Kenedy Co., 10 mi S of Norias, *Shinners 17812* (SMU); Sarita, Hwy 77, *Shaw & Allred 2016* (MO). Kleberg Co., Santa Gertrudis, S of Canelo Pasture, *Johnston 5439* (TEX); Kingsville, *Jones 29536* (POM). Maverick Co., Eagle Pass, 1883, *Havard s.n.* (US); 8.25 mi NW of Eagle Pass, *Cory 43853* (TEX). Nueces Co., Mustang Island, *Gillespie 206* (TEX). Refugio Co., Tivoli, *Palmer 9248* (MO). Starr Co., Rio Grande City, *Hanson 340* (MIN); 3 mi N of Roma, *Shinners 17712* (SMU); 9 mi N of Roma, *Alvarez et al. 8038* (TEX). Val Verde Co., Mill Canyon near Lantry, *Parks et al. 271* (LL, SMU). Victoria Co., Victoria, 1900, *Eggert s.n.* (MIN). Webb Co., Rio Grande near Laredo, *Mackenzie 30* (MO, NY); Laredo, *Jones 29534* (POM), *Reverchon 3757A* (DAO), *3757* (MO), *Palmer 11290* (MO); 3 mi S of Laredo, *Correll 20766* (LL); Casa Blanca Lake, 6 mi E of Laredo, *Villarreal 46* (OSH, TTC). Willacy Co., Yturria Station, *Runyon 1844* (WIS). Wilson Co., 6 mi SW of Floresville, *McCart 6314* (SMU). Zapata Co., 6 mi S of San Ygnacio, *Garcia 18* (TEX, 94% fertility).

Oenothera falfurriae is endemic to southeastern Texas. Its range is nearly the same as those of *O. cordata* and *O. mexicana.* When this species was first detected by Dietrich, the specimens were treated as hybrids between *O. grandis* and *O. laciniata,* because they are somewhat intermediate between these species. Seed samples collected by K. Allred and R. Shaw made it possible to cultivate plants of this species at the Botanical Institute in Düsseldorf, and it soon became obvious that these were by no means hybrids but instead represented an undescribed

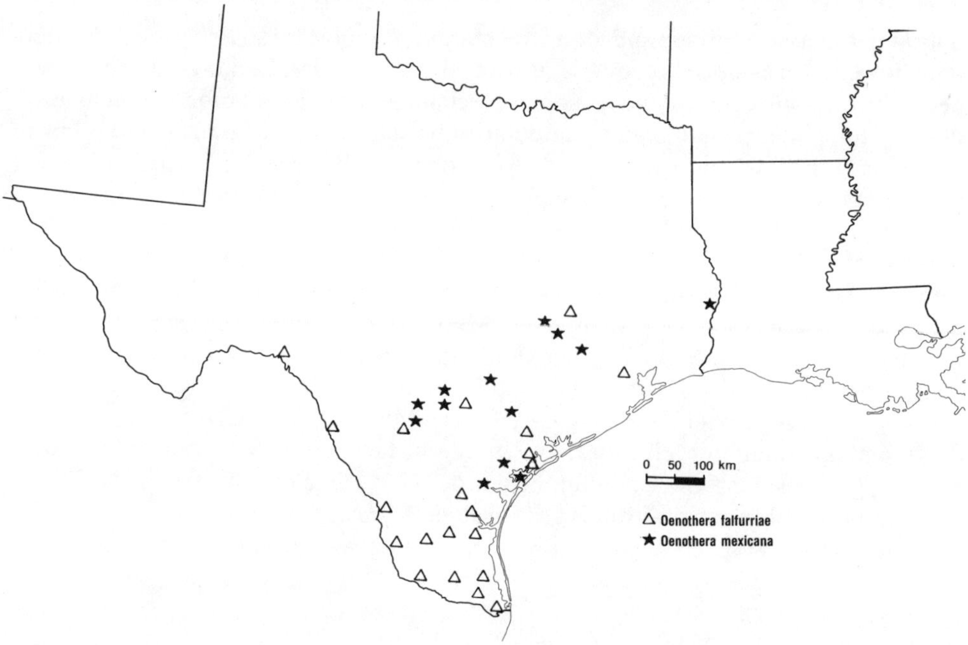

FIG. 6. Distribution of *Oenothera falfurriae* and *O. mexicana.*

bivalent-forming species. All plants examined formed 7_{II} in meiosis and had pollen stainability over 90%. No individuals grown from seed resembled either *O. laciniata* or *O. grandis.* The individual collection numbers of Allred and Shaw represent population samples of several plants from which seeds were taken and sown separately; *2016, 2020,* and *2021* contained *O. falfurriae* and *O. laciniata; 2018* contained *O. laciniata* and *O. mexicana.* Seeds taken from plants of *O. falfurriae* produced only individuals referrable to *O. falfurriae,* and seeds of *O. laciniata* produced only plants assigned to *O. laciniata.*

Oenothera falfurriae differs from *O. grandis* in its self-compatibility and smaller petals, which are intermediate in size between those of *O. laciniata* and those of *O. grandis.* Furthermore, the stigma in the closed mature buds are only slightly raised above the anthers, which suggests that self-pollination is common in this small-flowered species. Also, the shape of the buds is more or less oblong, in contrast to the lanceoloid buds of *O. grandis,* and the sepals in *O. falfurriae* are delicate and pressed together in bud, whereas in *O. grandis* they are often spreading, longer, and thicker in texture. Specimens with pollen fertility below 50% and similar morphology to *O. falfurriae* were treated as hybrids between *O. grandis* and *O. laciniata.*

Two extreme forms can be seen in *O. falfurriae;* one has deeply incised leaves, whereas the other almost always has subentire ones. Between these two extremes, however, transitional forms with more or less deeply toothed leaves occur in the same populations.

Oenothera falfurriae is a narrowly distributed and presumably relictual species. It appears to maintain itself from the other species of ser. *Raimannia* with which it grow sympatrically—*O. grandis, O. laciniata,* and *O. mexicana*—by possessing a unique plastome. Artificial crosses made by Dr. Behn at the Botanical Institute in

Düsseldorf show that crosses between *O. drummondii* or *O. humifusa* and *O. falfurriae* as the staminate parent produced pale seedlings that did not grow beyond the cotyledon stage before they died. Similarly, the seeds of crosses between *O. grandis* and *O. falfurriae* did not germinate at all (Behn, pers. comm.). Also, since crosses between *O. drummondii, O. humifusa,* or *O. grandis* and *O. laciniata* produce completely green and viable offspring, we can assume that similar crossing barriers exist between *O. falfurriae* and *O. laciniata;* this is the pattern of such relationships in *Oenothera* sect. *Oenothera* generally.

The trend in *O. falfurriae* toward smaller, autogamous flowers and the loss of self-incompatibility might represent an intermediate step toward the evolution of permanent structural heterozygotic species, such as *O. laciniata.* There is no evidence, however, that *O. falfurriae* occupies an evolutionarily intermediate position between *O. laciniata* and *O. grandis,* especially since it appears to have a different plastome from that which is present in *O. laciniata.*

8. Oenothera mexicana Spach, Nouv. Ann. Mus. Hist. Nat. 4: 347. 1835. *Oenothera sinuata* var. *hirsuta* Torrey & A. Gray, Fl. N. Amer. 1: 494. 1840. *Oenothera laciniata* var. *mexicana* (Spach) Small, Bull. Torrey Bot. Club 23: 173. 1896. *Raimannia mexicana* (Spach) Wooton & Standley, Contr. U.S. Natl. Herb. 19: 470. 1915.—Type: U.S.A. Texas: "between Brazos [River] and San Felipe [Austin]," 1833–34, *Drummond 75* (holotype: FI!; isotypes: BM! E! G! GH! GOET! K! NY! P! PH! TEX! W!).

Erect annual herbs, forming rosettes; main stem 1.5–4 (–6) dm long, up to 1 cm in diameter at base, simple or with arcuating lateral branches arising from the rosette, sometimes flushed with red, moderately to sparsely strigillose and densely long-villous. Rosette leaves 6–10 cm long, 1–2.5 cm wide, narrowly oblanceolate, deeply lobed, the lobes dentate, apex acute, gradually narrowed to the petiole; cauline leaves 3–7.5 cm long, 0.8–2 cm wide, narrowly oblanceolate to oblanceolate, deeply lobed to dentate, apex acute, base narrowly cuneate, sessile to short-petiolate; bracts 2–4 cm long, 0.7–1.2 cm wide, narrowly oblong or oblanceolate, dentate to subentire, apex acute, sessile, the uppermost ones erect, margins revolute; leaves and bracts grayish green, densely strigillose. Inflorescence often with lateral branches. Usually one flower per spike opening each day near sunset. Floral tube 2.3–2.8 cm long, ca. 1 mm in diameter, exclusively glandular-puberulent or scattered to very sparsely villous and sparsely to densely glandular-puberulent. Mature buds 3–4 mm in diameter at the base, oblong to narrowly ovoid or broadly ellipsoid. Sepals 0.5–1.2 cm long, green, sometimes red-flecked, pubescence like that of floral tube, sepal tips 0.5–2.5 mm long, erect and appressed in bud, strigillose. Petals 0.6–1.5 cm long, 0.7–1.7 cm wide, yellow to pale yellow, very broadly obovate, truncate to slightly retuse. Filaments 4–12 mm long; anthers 3–4 mm long, shedding pollen directly on the stigma at anthesis, pollen ca. 90–100% fertile. Ovary 0.8–1.2 cm long, ca. 1.5 mm in diameter, very densely strigillose. Style 2.7–4 cm long, the exserted part 0.4–1.3 cm long; stigma surrounded by the anthers at anthesis, lobes 1–3 mm long. Capsule 2.5–4.5 cm long, 2.5–3 mm in diameter, cylindrical, densely strigillose. Seeds 0.8–1.2 mm long, 0.3–0.5 mm in diameter, ellipsoid to broadly ellipsoid, brown, the surface pitted. Self-compatible and modally autogamous. Chromosome number: n = 7 (7_{II}* at meiotic metaphase I). Fig. 7.

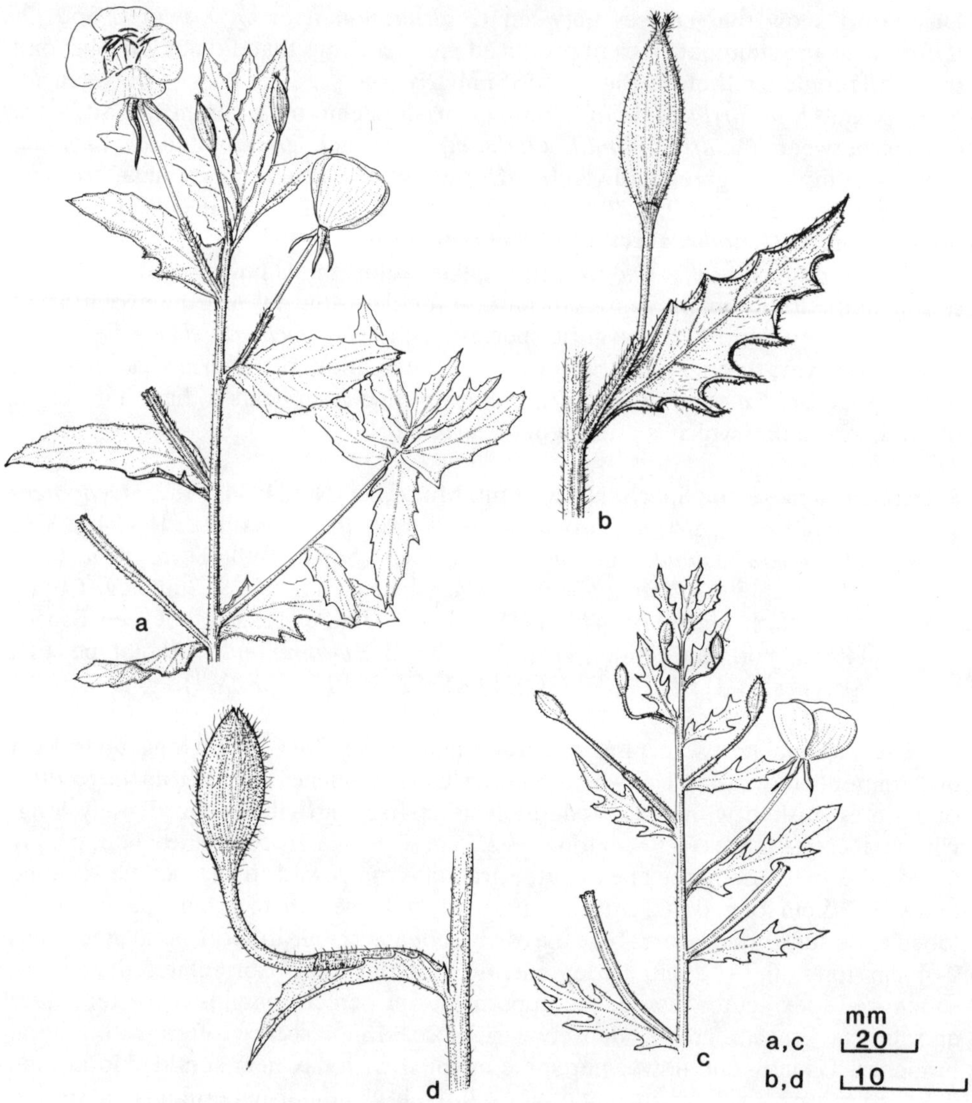

FIG. 7. *Oenothera mexicana* and *O. laciniata*. *O. mexicana* (*Ellstrand 5d,* cult. no. 79–110; MO): a. Inflorescence. b. Bud with bract. *O. laciniata* (*Baltzell 9674,* Florida, Hardee Co., Hickory Creek; cult. no. 79–156; MO; *Shaw & Allred 2016,* Texas, Kenedy Co., Sarita, cult. no. 79–127; MO): c. Inflorescence (*Baltzell*). d. Bud with bract (*Shaw & Allred*).

Phenology. Flowering in April and May.
Distribution (Fig. 6). Open sandy soils in southeastern Texas.

SPECIMENS FROM CULTIVATED PLANTS. U.S.A. TEXAS: Atascosa Co., 5 mi N of Leming, *Ellstrand 5c, 5d* [cult. no. 79–109*, 79–110*] (DUSS, M, MO), 1 mi W of RR 1784 on RR 536, *Ellstrand 6c, 6d, 6e* [cult. no. 79–112*, 79–113*, 79–114*] (DUSS, M, MO). De Witt Co., 0.5 mi S of Cuero on US 183, *Ellstrand 3b, 3c, 3d* [cult. no. 79–101*, 79–102*, 79–103*] (DUSS, M, MO). Gonzales Co., 3.2 mi N of jct. US 90A with Texas 80, 1977, *Ellstrand s.n.* [cult. no. 78–818*, 78–819*, 78–823*] (DUSS, M, MO).

Kenedy Co., 12.8 mi S of Riviera at Hwy 77, *Shaw & Allred 2018,* pro parte [cult. no. 80–321*, 322*, 323*, 325*] (DUSS, M, MO). Medina Co., Frontage road on I-35, *Ellstrand 4b, 4c, 4d* [cult. no. 79–105*, 106*, 106*] (DUSS, M, MO). Refugio Co., 4.5 mi NW of Hwy 77 near Woodsboro, *Hill 7264* [cult. no. 79–140*] (DUSS, M, MO). Washington Co., Yegua Creek Park, S side of Somerville Lake, *Ellstrand 7b, 7c, 7e* [cult. no. 79–115*, 116*, 117*] (DUSS, M, MO).

Additional Specimens Examined. U.S.A. Texas: Aransas Co., N of Rockport, *Whitehouse 19828* (MICH, NY, SMU, UC, US). Bexar Co., San Antonio, *Metz 578* (MASS). Brooks Co., N of Encino, *Bunyon 2647* (POM, TEX), *Lundell & Lundell 8838* (DS, GH, LL, MICH, NY, SMU, UC); 13.3 mi S jct. Hwys 281 & 285, Falfurrias, *Shaw & Allred 2021,* pro parte (MO). Burleson Co., Somerville, *Palmer 11698* (BM, GH, MO, NY, US); 12 mi SW of Caldwell, *Cory 51653* (MICH, NY, SMU). Frio Co., 2 mi E of Pearsall, *Gregory 283* (CAS, RSA, UC). Kenedy Co., 13.1 mi S of Riviera, *Mosquin 5566* (DAO, DS); 12.8 mi S of Riviera at Hwy 77, *Shaw & Allred 2018,* pro parte (MO). Medina Co., 2 mi SW of Devine, *Johnston et al. 3392* (MASS, RSA, SMU, TEX). Newton Co., 4 mi WSW of Newton, *Cory 57238* (CAS). Refugio Co., Woodsboro, *Hill 6732, 7264* (MO). San Patricio Co., 4 mi SE of Edroy on Hwy 9, *Whitehouse 18138* (MICH, NY, SMU, US). Waller Co., Hempstead, *Hall 202* (BM, F, GH, K, NA, NY, POM, US).

Oenothera mexicana has usually been considered to be a variety of *O. laciniata* (Torrey & Gray 1840; Small 1896, 1913) or a synonym of that species (Munz 1965). During our herbarium studies of ser. *Raimannia* in 1978 at the Missouri Botanical Garden, specimens that we would now regard as *O. mexicana* were always determined as *O. laciniata.* We now consider *O. mexicana* to be a distinct species, however, because pollen studies of the herbarium material always showed a pollen fertility of 90–100%. Cytological study of cultivated plants bore out our assumption that these plants are bivalent-forming (7_{II} at meiotic metaphase I).

A striking characteristic of *O. mexicana* is its thick spreading pubescence, which is much denser than that ever found in *O. grandis, O. falfurriae,* or *O. laciniata.* Also, the upper bracts are strictly erect and have revolute margins, whereas the other species of ser. *Raimannia* have conspicuously spreading, flat bracts. The upright bracts of *O. mexicana* appear to prevent the usual spreading of the ovary from the stems and the upward curve of the floral tube.

Oenothera mexicana, like the other bivalent-forming, autogamous species in subsect. *Raimannia, O. falfurriae,* has a limited geographical range. The reasons for the narrow distribution of these species are unclear. Despite its limited range, *O. mexicana* may have been involved in the formation of a form of the polymorphic *O. laciniata* in southeastern Texas, as is discussed further under that species.

9. Oenothera laciniata Hill, Veg. syst. 12, appendix: 64, *pl. 10.* 1767. *Raimannia laciniata* (Hill) Rose ex Britton & A. Brown, Ill. fl. n. U.S., ed. 2, 2: 59. 1913.—Type: We did not see authentic material and here select illustration *pl. 10* as the lectotype (Fig. 8). Plants for this illustration were obtained from "Caroliniana".

Oenothera sinuata Linnaeus, Mant. 2: 228. 1771. *Onagra sinuata* (Linnaeus) Moench, Methodus 676. 1794.—Type: Herb. Linné, *484.6* (LINN!).

Oenothera repanda Medikus, Act. Acad. Theod. Palat. 3: 198, *t. 8.* 1775.—Type: No authentic material seen, but *t. 8* clearly represents *Oenothera laciniata.*

Oenothera minima Pursh, Fl. Amer. sept. 1: 262, *t. 15.* 1814. *Oenothera sinuata* var. *minima* (Pursh) Nuttall, Gen. N. Amer. pl. 1: 245. 1818.—Type: U.S.A. Georgia: in barren pine-woods, *Enslen* (not located).

FIG. 8. Photograph of lectotype of *Oenothera laciniata*.

Oenothera laciniata f. *integrifolia* Jansen & Kloos in Kloos, Nederl. Kruidk. Arch. 1921: 99. 1922.—TYPE: THE NETHERLANDS. Leeuwen, Wamel, near a mill, 13 Aug 1917, *Kloos & Schmijtz s.n.* (lectotype, here designated: L!).

Erect to procumbent annual or short-lived perennial herbs, usually forming rosettes; stems 0.5–5 (in cultivation up to 10) dm long, green or flushed with red, simple to much-branched, sparsely to moderately strigillose and villous or only strigillose, in the region of the inflorescence often also glandular-puberulent. Rosette leaves 4–15 cm long, 1–3 cm wide, linear-oblanceolate to narrowly oblanceolate, occasionally lanceolate, deeply lobed to dentate; cauline leaves 2–10 cm long, 0.5–3.5 cm wide, narrowly oblanceolate to oblanceolate or narrowly oblong to narrowly elliptic, narrowly cuneate at base; bracts 2–7 cm long, 0.8–3 cm wide, spreading, narrowly oblong to narrowly ovate, flat, obtuse to narrowly cuneate at base; all leaves and bracts green, strigillose and villous, often also glandular-puberulent. Inflorescence lax, usually with lateral branches. One flower per spike opening each day near sunset. Floral tube 1.2–3.5 cm long, ca. 1 mm in diameter, yellowish often flushed with red, densely to sparsely villous and glandular-puberulent, sometimes also strigillose. Mature buds 2.5–5 mm in diameter at the base, narrowly oblong to narrowly ovoid, the older buds curved upward. Sepals 0.5–1.5 cm long, green to yellowish, often flushed with red and with a marginal red stripe, rarely flecked with red, sepal tips 0.3–3 mm long, usually spreading in bud, strigillose to villous. Petals 0.5–2.2 cm long, 0.7–2 cm wide, yellow to pale yellow, very broadly obovate, truncate to emarginate. Filaments 3–14 mm long; anthers 2–6 mm long, shedding pollen directly on the stigma at anthesis, pollen ca. 50% fertile. Style 2–5 cm long, the exserted part 0.3–1.4 cm long; stigma surrounded by the anthers at anthesis, lobes 2.5–5 mm long. Ovary 1–2.3 cm long, ca. 1.5 mm in diameter, strigillose or also villous, sometimes with a few glandular hairs. Capsule 2–5 cm long, 0.2–0.4 cm in diameter, cylindrical. Seeds 0.9–1.8 mm long, 0.4–0.9 mm in diameter, ellipsoid to suborbicular, brown to dark brown, the surface pitted. Autogamous, permanent structural heterozygote. Chromosome number: $n = 7$ ($\odot 14^*$ chromosomes at meiotic metaphase I). Fig. 7.

Phenology. Flowering from February to October along the Atlantic coast of the United States, from April to September elsewhere.

Distribution (Fig. 9). Open, usually sandy sites, primarily disturbed habitats, from North Dakota south to Kansas, Oklahoma, and Texas east to the Atlantic coast, introduced in California and in many parts of the world. It is puzzling that *O. laciniata* has been collected only once in Mexico, adjacent to Texas. We see no reason that the range should terminate abruptly as shown on the map, and expect that, if sought out, it will be found to occur more widely, at least in Tamaulipas.

Oenothera laciniata has been reported in the literature from outside of North America as follows: Australia (Bailey 1913, as *O. sinuata*); Austria (Rostanski & Forstner 1982); Belgium (Jean 1975, as *O. sinuata*); British Isles (Kent 1980; Rostanski 1982; Rostanski & Ellis 1979); France (Linder 1965, as *O. sinuata*); Germany (Thellung & Zimmermann 1916, as *O. humifusa*); Italy (Pignatti 1982, as *O. sinuata*); Japan (Asai 1973); Paraguay (Dietrich 1977); and South Africa (Guillarmod 1971; Ross 1972).

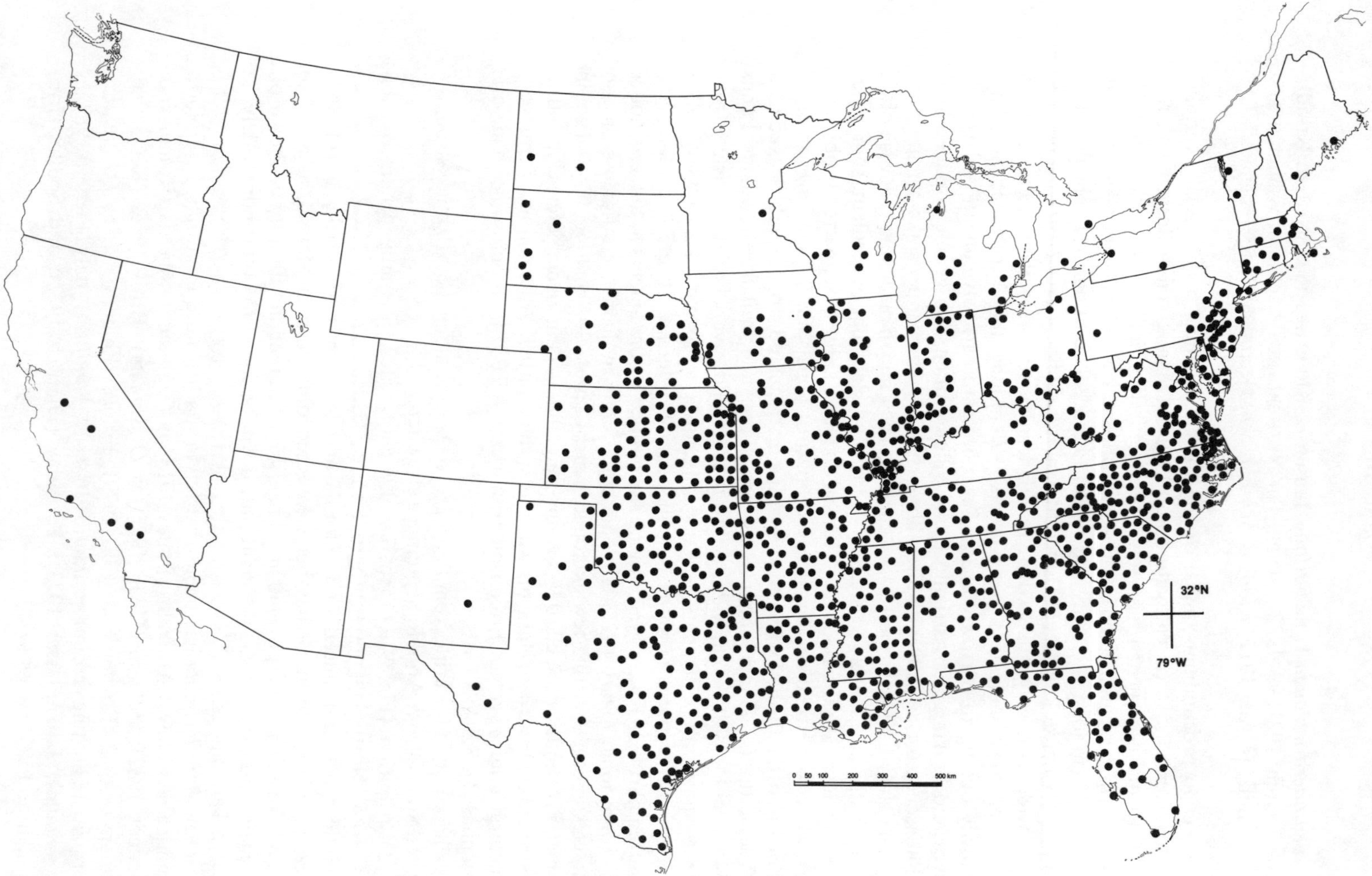

FIG. 9. Distribution of *Oenothera laciniata*.

SPECIMENS FROM CULTIVATED PLANTS: CANADA. Without further locality, *Kew 210-14* [cult. no. 74-314-20*] (DUSS, M, MO).——JAPAN. Kyushu, Sendai City, *Boufford & Mitsuta 19887* [cult. no. 78-240*] (DUSS, M, MO).——U.S.A. ALABAMA. Baldwin Co., betw. Mobile and Spanish Fort, *Kral 55663* [cult. no. 79–85*] (DUSS, M, MO).—ARKANSAS: Logan Co., Hwy 23, *Carter 277* [cult. no. 74–314–1*] (DUSS, M, MO). St. Francis Co., Forest City, *Hecht 21* [cult. no. 73–303*] (DUSS, M, MO). Washington Co., Fayetteville, 1974, *Smith s.n.* [cult. no. 77–585] (DUSS, M, MO).—FLORIDA: De Soto Co., Arcadia, *Baltzell 9602* [cult. no. 79–151*, 79–155*] (DUSS, M, MO); 6 mi NW Arcadia, *Baltzell 9800* [cult. no. 79–153*] (DUSS, M, MO); 5 mi SW Gardner, *Baltzell 6087* [cult. no. 79–154*] (DUSS, M, MO). Duval Co., Jacksonville, *Cleland 424* [cult. no. 73–304*] (DUSS, M, MO). Hardee Co., 6 mi SW Zolfo Springs, *Baltzell 9674* [cult. no. 79–156*] (DUSS, M, MO). Hernando Co., Nobelton, *Baltzell 8582* [cult. no. 79–157*] (DUSS, M, MO); 1.5 mi NNW Richlo, *Baltzell 9886* [cult. no. 79–158*] (DUSS, M, MO). Lake Co., 0.5 mi E Leesburg, *Baltzell 8475* [cult. no. 79–160*] (DUSS, M, MO). Leon Co., Tallahassee, *Kew 210–02* [cult. no. 76–768*] (DUSS, M, MO). Levy Co., 2 mi NE Cedar Key, *Baltzell 8505* [cult. no. 79–150*] (DUSS, M, MO); 2.9 mi NE Cedar Key, *Baltzell 8512* [cult. no. 79–162*] (DUSS, M, MO). Osceola Co., 10.5 mi W Kissimee, *Baltzell 9789* [cult. no. 79–149*] (DUSS, M, MO). Pasco Co., 7.5 mi NE Zephyrhills, *Baltzell 9778* [cult. no. 79–148*, 79–164*] (DUSS, M, MO), *Baltzell 9773* [cult. no. 79–165*] (DUSS, M, MO). Putnam Co., E Palatka, *Smith 3465* [cult. no. 78–811*] (DUSS, M, MO). St. Lucie Co., 0.5 mi N Ft. Pierce, *Baltzell 8338* (DUSS, M, MO).—GEORGIA: Oconee Co., 1 mi W Watkinsville, *Jones & Jones 22532* [cult. no. 77–592*] (DUSS, M, MO). Paulding Co., 6 mi SE Dallas, 1968, *Hoff s.n.* [cult. no. 74–314–9*] (DUSS, M, MO); 8 mi S Dallas, 1968, *Hoff s.n.* [cult. no. 74–314–8*] (DUSS, M, MO). South Cobb Co., without further locality, 1974, *Hoff s.n.* [cult. no. 77–596*] (DUSS, M, MO). Worth Co., 4 mi NW Sylvester, 1974, *Hoff s.n.* (DUSS, M, MO).—KANSAS: Barber Co., County State Lake, *Brooks 6646* [cult. no. 77–580*] (DUSS, M, MO). Comanche Co., 4 mi E Protection, 1974, *Brooks s.n.* [cult. no. 77–573*] (DUSS, M, MO). Edwards Co., 4 mi N Fellsburg, *Brooks 6314* [cult. no. 77–574*] (DUSS, M, MO); 5 mi E Lewis, *Brooks 6324* [cult. no. 77–575*] (DUSS, M, MO). Harper Co., Runnymede, *Stephens 79083* [cult. no. 77–581*, 77–582*] (DUSS, M, MO). Kingman Co., Kingman County Lake, *Brooks 6700* [cult. no. 77–576*] (DUSS, M, MO). Kiowa Co., 2 mi E Brenham, *Brooks 6621* [cult. no. 77–579*] (DUSS, M, MO). Reno Co., 1.5 mi W Sylvia, *Brooks 6307* [cult. no. 77–577*] (DUSS, M, MO); Medora, *Brooks 5829* [cult. no. 77–578*] (DUSS, M, MO). Riley Co., *Anderson 3670* [cult. no. 75–1508*] (DUSS, M, MO). Stafford Co., 4 mi S St. John, *Brooks 6310* [cult. no. 75–1480*] (DUSS, M, MO).—LOUISIANA: Orleans Pa., without further locality, *Lieu 2544* [cult. no. 78–812*] (DUSS, M, MO).—MISSISSIPPI: Forrest Co., McLaurin, *Rogers 6668-A* [cult. no. 78–806*] (DUSS, M, MO). Harrison Co., Biloxi peninsula, 1975, *Argnelles s.n.* [cult. no. 77–603*] (DUSS, M, MO).—MISSOURI: Missouri Botanical Garden, St. Louis, spontaneous, 1974, *Raven s.n.* [cult. no. 77–602*] (DUSS, M, MO).—NORTH CAROLINA: Without further locality, *Kew 209–12* [cult. no. 77–597*] (DUSS, M, MO). Harnett Co., 11 mi N Millington, *Broome 854* [cult. no. 77–600*] (DUSS, M, MO), *Kew 209–10* [cult. no. 75–1460*] (DUSS, M, MO).—TEXAS: Angelina Co., without further locality, 1974, *Hoff s.n.* [cult. no. 77–591*] (DUSS, M, MO). Atascosa Co., without further locality, *Ellstrand 6g, 6i* [cult. no. 79–143*, 79–144*] (DUSS, M, MO). Brooks Co., Falfurrias, *Shaw & Allred 2020,* pro parte [cult. no. 79–136*, 79–147*, 80–310*, 80–312*, 80–313*, 80–315*, 80–317*] (DUSS, M, MO), *Shaw & Allred 2021,* pro parte [cult. no. 79–139*, 80–327*, 80–328*, 80–329*, 80–330*, 80–331*] (DUSS, M, MO). De Witt Co., Cuero, *Ellstrand 3g, 3h* [cult. no. 79–141*, 79–142*] (DUSS, M, MO). Kenedy Co., Sarita, *Shaw & Allred 2016,* pro parte [cult. no. 79–121*, 79–122*, 79–123*, 79–124*, 79–125*, 79–126*, 79–127*, 79–128*, 79–130*, 79–131*, 80–332*, 80–333*, 80–334*, 80–335*, 80–336*, 80–337*, 80–338*, 80–339*, 80–340*, 80–342*, 80–343*] (DUSS, M, MO); 12.8 mi S Riviera, Hwy 77, *Shaw & Allred 2018,* pro parte [cult. no. 79–132*, 80–318*, 80–319*, 80–324*] (DUSS, M, MO). Kleberg Co., N Riviera, 1972, *Hoff s.n.* [cult. no. 77–611*] (DUSS, M, MO). Nacogdoches Co., Nacogdoches, 1968, *Hoff s.n.* [cult. no. 74–314–5*] (DUSS, M, MO). Nueces Co., Padre Island, 1974, *Jones s.n.* [cult. no. 78–247*] (DUSS, M, MO). Pottawatomie Co., 5 mi E Manhattan, *Anderson 3680* [cult. no. 75–1449*] (DUSS, M, MO). Van Zandt Co., 15 mi E Kaufman, 1968, *Hoff s.n.* [cult. no. 74–314–10*] (DUSS, M, MO). Washington Co., Hwy 36 near county line with Burleson Co., *Shaw & Allred 2007* [cult. no. 79–145*, 80–344*, 80–345*, 80–346*] (DUSS, M, MO).—VIRGINIA: City of Virginia Beach, Virginia Beach, 1974, *Straley s.n.* [cult. no. 75–1510*, 77–612*] (DUSS, M, MO).——FRANCE. Hort. Bot. Paris, 1837 (FI-W).——GERMANY: Botanical Garden Erlangen, 1779, *Schreber s.n.* (M); Erlangen, 1780, 1781, and 1784, *Schreber s.n.* (M) (as *O. sinuata*); 1819, h R P (DS); Botanical Garden Munich, 1821 (M); Botanical Garden Berlin, 1820–32, *Schlechtendal s.n.* (HAL-43439).

Representative Specimens. Canada. Ontario: Elgin Co., Aug 1929, *Howitt s.n.* (DAO). Norfolk Co., 1951, *Landon s.n.* (DAO).——Mexico. Tamaulipas: 12 mi S of Nuevo Laredo, *Saenz 65* (DS, DUKE, SMU).——U.S.A. Alabama: Autauga Co., *Shinners 29752* (SMU). Baldwin Co., *McDaniel 4251* (ALU, VDB). Barbour Co., *Kral 28167* (VDB). Blount Co., *Kral & Demaree 30720* (VDB). Bullock Co., *Mann 66* (ALU). Calhoun Co., *Dean 80* (NCU). Chambers Co., *Rutland 145* (AUA). Cherokee Co., *Kral 30310* (C, SMU). Clay Co., *Rutland 2484* (AUA). Cleburne Co., *Rutland 2774* (AUA). Colbert Co., *Isely 3156* (ISC). Coosa Co., *Rutland 2300* (AUA). Dale Co., 1966, *Logan s.n.* (AUA). Dallas Co., *Callaway 54* (ALU). DeKalb Co., *Straley 888* (VPI). Elmore Co., *Andrews C-36* (NCU). Escambia Co., *Ahles 7114* (ILL). Fayette Co., *Cooley & Brass 3567* (GH, USF). Geneva Co., *Hardin & Duncan 14940* (GA). Greene Co., *Harper 4545* (ALU, NCU). Hale Co., *Williams 233* (ALU, NCU). Houston Co., *Kral 46185* (VDB). Jackson Co., 1930, *Porter s.n.* (GH). Jefferson Co., 1907, *Setchell s.n.* (UC). Lamar Co., *Clark 10789* (NCU). Lauderdale Co., 1974, *Hoffman s.n.* (MIN). Lee Co., 1897, *Earle & Baker s.n.* (F, KSC, MIN, MO, KY). Limestone Co., *Gillespie & Dodd 907* (FSU, GH). Madison Co., 1968, *White s.n.* (NCU). Marion Co., *Shinners 31210* (SMU). Marshall Co., *Kral 45955* (VDB). Mobile Co., *Deramus D261* (ALU, MO). Monroe Co., *Uttal 11019* (NCU, VPI). Montgomery Co., *Justice 192* (CU). Morgan Co., 1891, *Shimek s.n.* (F, ISC). Pickens Co., *Kral 45322* (VDB). Pike Co., *Koehler 48* (NCU). Randolph Co., *Rutland 205* (AUA). St. Clair Co., *Little 164* (AUA). Shelby Co., *Clark 13239* (NCU). Talladega Co., *Clark 4162* (NCU). Tallapoosa Co., *Rutland 2059* (AUA). Tuscaloosa Co., 1964, *Williams s.n.* (ALU).—Arkansas: Arkansas Co., *Demaree 21041* (BH, MO, NY, POM, SMU). Ashley Co., *Demaree 20796A* (GH, ISC, MIN, MO, NY, POM, SMU). Baxter Co., *Demaree 23550* (MO, NY, POM). Benton Co., *Demaree 52527* (DS, SMU). Calhoun Co., *Reddin 25* (UARK). Carroll Co., *Bush 14871* (BHO, G). Clark Co., *Youree 102* (UARK). Clay Co., *Demaree 28864* (ISC, SMU, USFS). Cleburne Co., *Demaree 30539* (KANU, ISC, RSA, SMU, TEX). Cleveland Co., *Demaree 17143* (MO, NY, POM, US). Columbia Co., 1899, *Hurth s.n.* (NY). Conway Co., *Demaree 21351* (DAO, POM). Craighead Co., *Demaree 3421* (MICH, MO, SMU, UARK, WIS). Crawford Co., *Isely 6386* (ISC, US). Crittenden Co., *Demaree 15259* (DAO, DS, GH, MO, NY, OKL, POM, WIS). Dallas Co., 1969, *Wright s.n.* (UARK). Desha Co., *Demaree 21032* (NY, POM). Drew Co., *Demaree 14964* (DAO, ILL, MIN, MO, NY, OS, POM, TENN, WIS) . Faulkner Co., *Demaree 5862* (NY). Franklin Co., *Demaree 20887* (MO, NY, SMU). Fulton Co., *Demaree 25998* (KANU, ISC, MIN, RSA, SMU, TEX). Garland Co., *Demaree 36471* (GA, GH, KANU, MSC, NCU, OKL, SMU). Grant Co., *Demaree 14980* (POM, SMU). Hempstead Co., *Demaree 42266* (GH, OKL, SMU). Hot Spring Co., *Demaree 14857* (GH, MIN, MO, NY, OKL, POM, SMU, UMO). Izard Co., *Demaree 17032* (MO, NY, POM). Jefferson Co., *Demaree 19084* (MO, NY, POM). Johnson Co., *Demaree 51655* (DS, SMU). Lafayette Co., *Shinners 28543* (SMU). Lawrence Co., *Demaree 30448* (DS, KANU, ISC, MIN, OKL, SMU, TEX, WVA). Lee Co., *Griffen & Demaree 17* (DS, DUKE, MIN, MO, SMU, WIS). Lincoln Co., *Demaree 19167* (MO, NY, POM, SMU). Little River Co., *Raper 27* (UARK). Logan Co., *Demaree 17732* (GH, ISC, MIN, NY, OKL, POM, SMU, UC, USF). Lonoke Co., *Demaree 17592* (GH, ISC, MIN, MO, NY, OKL, POM, SMU, UC). Madison Co., *Fritts 51* (UARK). Marion Co., *Demaree 47963* (DS). Miller Co., *Demaree 40820* (SMU, UC). Montgomery Co., *Demaree 45464* (SMU, TENN, VDB, VPI). Nevada Co., *Tucker 3838* (NCU). Newton Co., *Thompson 302* (SMS, UARK). Ouachita Co., *Demaree 48167* (DS, VDB). Perry Co., *Wright 69107* (UARK). Phillips Co., *Demaree 55989* (DS). Pike Co., *Demaree 55330* (DS). Poinsett Co., *Marsh 2031* (UARK). Polk Co., *Hopkins et al. 606* (GH, MO, OKL, OKLA, UC). Pope Co., *Merrill 272* (ILL, MIN, OKLA). Prairie Co., *Demaree 37762* (ASTC, GH, KANU, NCU, OKL, SMU). Pulaski Co., *Demaree 17523* (BH, MO, NY, POM). St. Francis Co., *Demaree 15153* (MO, NY, OKL, SMU, USF). Saline Co., *Demaree 36465* (BH, DUKE, GA, GH, KANU, NCU, OKL, RSA, SMU, UNM, USC, VDB, VPI). Sebastian Co., *Measeles 17* (NCU, UARK). Sharp Co., *Demaree 26091* (ISC, KANU, MIN, RSA, SMU, TEX). Stone Co., *Demaree 55784* (DS, NCU, SMU, TENN, VDB). Union Co., *Gillett 5933* (DAO, UARK). Washington Co., 1958, *Finklestein s.n.* (KANU, NCU). Woodruff Co., *Huenefeld 75* (UARK). Yell Co., *Demaree 21266* (MO, NY, POM).—Connecticut: Fairfield Co., *Eames 5952* (CONN). Hartford Co., 8 Jul 1898, *Smith s.n.* (CONN). Litchfield Co., *Eames & Austin 8302* (CONN). New Haven Co., *Harger 4543* (CONN, PH). New London Co., 14 Aug 1939, *Jansson s.n.* (CONN).—Delaware: Kent Co., *Tatnall 3675* (PH). New Castle Co., 30 Jun 1878, *Commons s.n.* (PH). Sussex Co., *True 128* (PENN).—District of Columbia: U.S. National Arboretum, *Meyer & Mazzeo 1089* (ISC, UC, US, WVA).—Florida: Alachua Co., *Wiggins & Wiggins 19543* (FLAS). Bay Co., 4 Apr 1933, *Bailey & Bailey s.n.* (BH). Bradford Co., 15 May 1940, *West & Arnold s.n.* (FLAS). Citrus Co., *Schallert 3686* (ILL, KANU, NCU, SMU). Clay Co., 22 Mar 1884, *Rau s.n.* (PH). Columbia Co., *Bedistha G51* (F,

FLAS, MO). Dade Co., 1898, *Hitchcock s.n.* (F). Dixie Co., *Godfrey 56459* (FSU, GH, USF). Duval Co., *Curtiss 905* (AC, F, FLAS, GA, GH, IA, KANU, MIN, MO, MSC, NY, PH, UC). Escambia Co., *Gentry 6* (WIS). Flagler Co., *West & Arnold 171* (FLAS). Franklin Co., *Demaree 50389* (DS, SMS, SMU). Gadsden Co., 18 Apr 1930, *Kincaid & West s.n.* (FLAS). Gulf Co., 1858, *Canby s.n.* (NY). Hernando Co., *Moldenke 1074* (DUKE, ILL, MO, NY, PENN). Highlands Co., *Brass 15221* (GH, US). Hillsborough Co., *Lakela 25145* (DS, NY, USF). Jackson Co., 5 Apr 1944, *Knight s.n.* (FLAS). Jefferson Co., *Godfrey 58048* (FSU). Lake Co., *Nash 225* (C, CU, E, F, G, GH, MICH, MIN, MO, MSC, NY, PH, UC, US). Lee Co., *Brumbach 8303* (GH, MICH, NCU, NY). Leon Co., *Godfrey 53005* (DUKE, GH, NY, VPI). Levy Co., *Wiggins 19331* (DS, GA). Madison Co., *Kral 4191* (FSU, GH, VPI). Manatee Co., *Genelle & Fleming 1955* (USF). Marion Co., 16 Apr 1936, *Perkins s.n.* (CU). Martin Co., 1917, *Atwood s.n.* (CU). Monroe Co., *Aredood 216* (USF). Nassau Co., *Boufford 5168* (MASS). Okaloosa Co., *Demaree 57760a* (GA, OS, VPI). Orange Co., *Fredholm 5400* (GH, ISC, MO, US). Osceola Co., *Cooley et al. 12632* (FLAS, USF). Pasco Co., *Cooley & Monachino 5672* (FSU). Pinellas Co., *Genelle & Fleming 2852* (USF). Polk Co., *Demaree 49448* (DS, NCU, RSA, SMU). Putnam Co., 17 Mar 1940, *Laessle s.n.* (FLAS). St. Johns Co., 3 Mar 1918, *Patterson s.n.* (CM). Santa Rosa Co., 1 May 1960, *Turner s.n.* (ALU). Seminole Co., *Cooley et al. 7545* (NCU, USF). Sumter Co., *Baltzell 353* (FLAS). Swannee Co., 17 Mar 1939, *Arnold s.n.* (FLAS). Taylor Co., *Godfrey 69380* (FSU). Volusia Co., *Hood 901* (FLAS). Walton Co., 9 May 1938, *Hume s.n.* (FLAS). Washington Co., 23 Apr 1937, *Hodges s.n.* (FLAS).—Georgia: Bacon Co., 17 Apr 1970, *Lipscomb & Erwin s.n.* (NCU). Baker Co., 19 Mar 1965, *Netting s.n.* (CM). Baldwin Co., 28 Apr 1972, *Pierce s.n.* (NCU). Barrow Co., *Thomas 1002* (ALU, NCU). Bartow Co., *Greear 66166* (GA, NCU). Berrien Co., *Bozeman 4765* (NCU). Bibb Co., *Benedict 3321* (NDA). Bleckley Co., *Hardin & Duncan 16034* (GA). Brooks Co., *Faircloth 3354* (GA, NCU). Bulloch Co., *Craig & Craig 3377* (POM). Chatham Co., May 1928, *Richards s.n.* (CU). Chattahoochee Co., *Lazor 3194* (FSU, NCU). Cherokee Co., *Duncan 8215* (GA, MO). Clarke Co., *Demaree 52017* (DS, MASS, SMU, VDB). Clay Co., *Williams 48* (AUA). Colquitt Co., *Demaree 49533* (MASS, NCU, VDB). Columbia Co., *Duncan 9725* (GA). Dougherty Co., 28 Apr 1975, *Drummond s.n.* (NCU). Decatur Co., *Godfrey 69338* (FSU, NCU). DeKalb Co., *Perry & Myers 968* (GH, NY). Floyd Co., *Jones 238* (VDB). Fulton Co., *Darby 228* (FSU). Glynn Co., *Cronquist 4274* (GA, GH, NY, PH, SMU, US). Grady Co., *Godfrey 65763* (FSU). Habersham Co., 7 May 1926, *Phillips s.n.* (CU). Hall Co., *Duncan & Adams 19791* (GA). Haralson Co., July 1900, *Way s.n.* (US). Harris Co., *Miller et al. 20725* (GA). Houston Co., *Cibulka 76* (VDB). Jasper Co., *Porter 1846* (PH). Jeff Davis Co., 17 Apr 1970, *Lipscomb & Erwin s.n.* (NCU). Laurens Co., 9 Apr 1970, *Lipscomb & Erwin s.n.* (NCU). Liberty Co., 11 Apr 1944, *Grimm s.n.* (CM). Lincoln Co., *Cronquist 4302* (GA, GH, NY, SMU). Long Co., *Bozeman 2692* (NCU). Lowndes Co., *Faircloth 1907* (GA, MO, NCU). McDuffie Co., *Bartlett 925* (IND, MICH, POM). McIntosh Co., *Moldenke & Moldenke 26418* (CU, TEX). Mitchell Co., *Bozeman 5162* (NCU). Murray Co., *Bowers & Morton 43543* (TENN, VDB, UPI). Muscogee Co., *Shehane 32* (AUA). Oconee Co., *Williams 68* (GA). Oglethorpe Co., 20 Apr 1967, *Harima s.n.* (ALU, VDB). Rabun Co., *DuMond 1212* (G). Richmond Co., 23 Apr 1904, *Guthbert s.n.* (YU). Rockdale Co., *Johnson 27* (GA). Screven Co., *Ahles 54327* (NCU). Seminole Co., *Kral & Knott 4243* (FSU). Sumter Co., *Demaree 49550* (DS, NCU, SMU, TENN, VDB). Tatnall Co., *Cronquist 4951* (GA, GH, MO, NY, PH, SMU, US). Taylor Co., 20 May 1972, *Register s.n.* (NCU). Thomas Co., *Oosting 95* (DUKE). Tift Co., 13 Apr 1926, *Haltern s.n.* (GA). Toombs Co., 12 May 1972, *Stinson s.n.* (NCU). Towns Co., *Massey & Massey 3698* (NCU). Walker Co., *Hunt 6* (GA). Ware Co., *Rush 150* (AUA). Wheeler Co., 9 Apr 1970, *Lipscomb & Erwin s.n.* (NCU). Wilkes Co., *Fitzgerald 287* (GA). Wilkinson Co., 17 Apr 1970, *Lipscomb & Erwin s.n.* (NCU).—Illinois: Adams Co., *Brinker 3619* (ILL, ILLS). Alexander Co., *Jones 12020* (ILL, NY, SMU). Calhoun Co., *Evers 94684* (ILLS). Carroll Co., *Jones 17300* (ILL). Cass Co., June 1901, *McDonald s.n.* (F, ILL). Champaign Co., *Ahles 5953* (ILL). Clark Co., *Evers 46781* (ILLS). Coles Co., *Evers 53558* (ILLS). Cook Co., 28 Jul 1907, *Greenman s.n.* (MINN, MO). Crawford Co., *Pepoon & Barrett 5424* (ILLS). Cumberland Co., 15 Jul 1950, *Myers s.n.* (WVA). Fulton Co., *Chase 10571* (ILL). Gallatin Co., *Ahles 2235* (ILL). Greene Co., *Evers 22994* (ILLS, WIS). Hamilton Co., *Evers 110397* (ILLS). Hancock Co., *Evers 96058* (ILLS). Henderson Co., *Evers 18355* (ILLS). Jackson Co., 9 May 1902, *Gleason s.n.* (GH, ILL, IND, ISC). Jefferson Co., *Ahles 2350* (ILL). Jersey Co., *Evers 108497* (ILLS). Jo Daviess Co., 7 Jun 1945, *Jones s.n.* (ILL). Johnson Co., *Evers 23736* (ILLS). Kankakee Co., *Jones 17278* (ILL). Lawrence Co., 21 Aug 1946, *Sivert s.n.* (ILL, MIN). Macon Co., 25 Jul 1942, *Mills s.n.* (BHO, F, ILL). Macoupin Co., *Evers 3841* (ILLS). Madison Co., *Evers 25075* (ILLS). Marshall Co., *Evers 33716* (ILLS). Mason Co., *Chase 10393* (F, ILL). Massac Co., *Evers 110320* (ILLS). Menard Co., *Evers 64523* (ILLS). Monroe Co.,

Pfeifer 2375 (CONN). Morgan Co., *Evers 69551* (ILLS). Ogle Co., *Ahles 4550* (ILL). Peoria Co., *Chase 8728* (IA, ILL, ILLS, KANU, ND, OKL, SMU, US). Perry Co., *Evers 87028* (ILLS). Pike Co., *Evers 99127* (ILLS). Pope Co., *Evers 82900* (ILLS). Pulaski Co., 22 May 1947, *Boewe s.n.* (ILLS). Putnam Co., *Chase 10451* (F, ILL, NY). Randolph Co., *Ahles 6123* (ILL). Richland Co., *Ridgway 2699* (POM). St. Clair Co., 2 Jun 1874, *Eggert s.n.* (CM, F, MO, ND, NY, OS, TEX, UC, US). Scott Co., *Evers 81141* (ILLS). Stark Co., *Chase 1888* (ILL). Tazewell Co., *Chase 13714* (ARIZ, DAO, ILL, KANU, OKL, SMU, UC). Union Co., *Fuller & Fisher 99* (F, ILLS). Vermilion Co., *Seigler 5051* (ILL). Wabash Co., *Henderson 67–778* (FSU, SMS, UMO, VDB). Washington Co., *Evers 23894* (ILLS). Wayne Co., *Walker 92* (ILL). White Co., *Ahles 2173* (ILL). Williamson Co., *McCree 799* (ILL, MO).—INDIANA: Daviess Co., Sep 1937, *Rhoades s.n.* (POM, TENN). Elkhart Co., *Deam 52325* (MIN). Gibson Co., *Deam 48755* (GH). Fountain Co., *Buser 2740* (ILL, IND). Greene Co., *Friesner 22133* (KANU, MICH, NY, OKL, OKLA, TEX, UC). Harrison Co., *Deam 35536* (IND). Jackson Co., *Deam 35695* (IND). Jefferson Co., *Butler 8904* (IND). Johnson Co., *Friesner 13588* (GA). Knox Co., *Deam 16996* (IND). Kosciusko Co., *Humbles 179* (IND). Lagrange Co., *Deam 36296* (IND). Lake Co., 27 Jul 1909, *Finger s.n.* (MIL). LaPorte Co., *Jones 22258* (ILL). Lawrence Co., *Deam 17288* (IND). Marion Co., *Friesner 13390* (DUKE, UC, WVA). Marshall Co., *Deam 31865* (IND). Martin Co., *Deam 11434* (IND). Monroe Co., *Smith 165* (IND). Newton Co., *Keil 7465* (OS). Owen Co., 13 Jun 1943, *Everly s.n.* (IND). Pike Co., *Deam 16963* (IND, NY). Posey Co., *Deam 950* (IND, MO, US). Pulaski Co., *Deam 49026* (CU, IND). St. Joseph Co., 6 Jul 1930, *Lyon s.n.* (MICH). Steuben Co., *Deam 59696* (IND). Sullivan Co., *McCoy 6003* (DAO, UMO). Tippecanoe Co., *Webster & Webster 7076* (DUKE, NCU). Tipton Co., *Ek 82* (BH, US). Vanderburgh Co., 12 Jul 1941, *Zelner s.n.* (IND). Vigo Co., *Deam 38738* (IND). White Co., Summer 1929, *Gordon s.n.* (OS).—IOWA: Clarke Co., *Bruggen 2353* (IA, UC). Clinton Co., *Cooperrider 2093* (IA, ILL, MIN, NCU). Dallas Co., *Maxon 7* (ISC). Davis Co., *Hayden 11400* (ISC). Delaware Co., *Rickey 526* (IA). Des Moines Co., *Davidson 2244* (IA, SMU). Fremont Co., *Fay 2751* (IA, UC). Jackson Co., *Cooperrider 4063* (IA). Johnson Co., 1 Jun 1955, *Davidson s.n.* (DAO, DS, IA, USF). Lee Co., *Davidson 456* (NCU). Linn Co., *Thorne et al. 14115* (IA). Louisa Co., *Davidson 4177* (IA). Mahaska Co., *Augustine 173* (ISC). Mills Co., *Morrill 446* (ISC). Muscatine Co., 3 Jul 1915, *Shimek s.n.* (GH, ISC). Polk Co., *Bruggen 1668* (IA). Pottawattamie Co., 7 Jun 1919, *Shimek s.n.* (ISC). Story Co., *Pohl 9360* (CAS). Warren Co., *Bruggen 1684a* (IA). Wayne Co., 11 Jun 1902, *Shimek s.n.* (ISC).—KANSAS: Allen Co., Sep 1869, *Hall s.n.* (F, ILL). Anderson Co., *Stephens 30554* (KANU). Atchison Co., *Blocker 159* (KSC). Barber Co., *Stephens 57699* (KANU). Barton Co., *Brooks 4352* (KANU). Bourbon Co., *Thompson 558* (KANU). Butter Co., *Stephens 4455* (KANU). Chase Co., *Stephens 4525* (KANU). Chautauqua Co., *Brooks 5496* (KANU). Cherokee Co., *Stephens 10461* (DS, KANU). Clark Co., *Brooks 6526* (KANU). Clay Co., 26 Jun 1899, *Schaffner s.n.* (OS). Cloud Co., *McGregor 24802* (KANU). Coffey Co., *Henderson 66–278* (KANU, MASS, SMS, UMO, VT). Comanche Co., *Brooks 6563* (KANU, MO). Cowley Co., *Koch 3791* (KANU, KSC, NCU, OKLA, UWL). Crawford Co., 11 May 1963, *Childress s.n.* (NCU). Dickinson Co., 8 Jun 1929, *Imler s.n.* (KANU). Douglas Co., *Horr E64* (CM, DAO, DUKE, F, FLAS, GH, ILL, IND, ISC, KANU, MIN, NY, OKL, PENN, TAES, TEX, UC, US, WIS, WVA). Edwards Co., *Brooks 6324* (MO). Ellis Co., 25 May 1936, *Bondy s.n.* (ARIZ, CAN, CAS, CU, DUKE, F, MIN, MO, NMC, OKL, OKLA, PENN, SMU, UARK, US, VDB, WIS, WVA). Ellsworth Co., *Wagenknecht 2471* (KANU). Emporia Co., 23 Jun 1960, *Weber s.n.* (UWL). Ford Co., *Brooks 6452* (KANU). Franklin Co., 22 May 1930, *Benson s.n.* (UMO). Geary Co., 28 May 1927, *Imler s.n.* (KANU). Graham Co., Summer 1941, *Chipman s.n.* (KSC). Grant Co., *Stephens 64887* (KANU). Greenwood Co., *McGregor 17213* (KANU). Harper Co., *Stephens 79083* (KANU, MO). Harvey Co., *Stephens 84487* (KANU). Hodgeman Co., *Brooks 6413* (KANU). Jefferson Co., May 1945, *Shoner s.n.* (KSC). Jewell Co., *McGregor 24861* (KANU). Johnson Co., *Brooks 9234* (KANU). Kearney Co., Aug 1949, *Browne s.n.* (KSC). Kingman Co., *Stephens 11123* (DS, KANU). Kiowa Co., *Brooks 6621* (MO). Labette Co, *Stephens 4214* (KANU). Leavenworth Co., *Mize s.n.* (KSC). Lincoln Co., *Stephens 64433* (KANU). Linn Co., *Rydberg & Imler 125* (KANU, KSC, NY). Lyon Co., *Brooks 9317* (KANU). McPherson Co., Jul 1887, *Rodin s.n.* (F, MIN). Miami Co., *Rohrer 28* (KSC). Mitchell Co., *McGregor 24811* (KANU). Montgomery Co., *Horr E143* (DUKE, OKL, OKLA). Morton Co., *Richards 2986* (KANU, SMU). Osage Co., *Brooks 9285* (KANU). Ottawa Co., *Stephens 30756* (KANU, NY). Pawnee Co., *Stephens 84903* (KANU). Phillips Co., *McGregor 24874* (KANU). Pottawatomie Co., *Anderson 3680* (FSU, MO). Pratt Co., *Stephens 47639* (KANU). Reno Co., *Brooks 5829* (MO). Republic Co., *Morley 72* (KANU). Rice Co., *Stephens 65233* (KANU). Riley Co., *Norton 158* (GH, KSC, MO, NMC, NY, US, YU). Rooks Co., 6 Jun 1887, *Bartholomew s.n.* (KSC). Russell Co., *McGregor 29012* (KANU). Sabine Co., *Raven & Gregory 19490* (DS, RSA, US).

Saline Co., *McGregor 28976* (KANU). Sedgewick Co., Jun 1903, *Poole s.n.* (GH). Shawnee Co., *Henderson 69–172* (FSU, VDB). Sheridan Co., *Weber 392* (KSC). Sherman Co., Jul 1894, *Smyth s.n.* (KSC). Stafford Co., *Ungar 769* (KANU, SMU). Sumner Co., *McGregor 16876* (KANU). Wabaunsee Co., *Stephens 3396* (KANU). Washington Co., *McGregor 4817* (GH, KANU, NY). Wilson Co., *Hulbert 3695* (B, KANU, KSC, OKLA). Woodson Co., *Lathrop 405* (MIN, NY, US). Wyandotte Co., *Richards 3246* (KANU).—Kentucky: Ballard Co., *Conrad 841* (MO). Bath Co., *Setser 675* (KY). Breathitt Co., *Barbour 19* (KY). Calloway Co., *Smith & Hodgdon 4085* (F, NY, US). Carlisle Co., *Hickok & Fuller 23* (NCU). Estill Co., *Demaree 55876* (DS, MASS, SMU, VDB). Fulton Co., *McInteer 1753* (KY). Henderson Co., 15 May 1918, *s.c. 2548* (UWM). Hickman Co., *Bailey 258* (BH, US). Jefferson Co., *Davies 498* (NCU). Lewis Co., *Braun 4498* (US). Livingston Co., *Windler & Windler 2460* (NCU). McCreary Co., *Braun 1423* (US). McLean Co., 25 May 1969, *Meijer & Conrad s.n.* (KY). Rowan Co., *Setser 494* (KY). Thomas Co., *Browne & Browne 6599* (KY). Trigg Co., *Ellis 64* (NCU). Warren Co., 15 Jun 1930, *Denniston s.n.* (WIS).—Louisiana: Acadia Pa., *Ellis 29* (LSU). Allen Pa., *Brown 18816* (LSU). Bienville Pa., *Demaree 20815* (BH, GH, ISC, MIN, MO, NY, POM). Bossier Pa., *Correll & Correll 10073* (DUKE, GH). Calcasieu Pa., *Brown 8952* (LSU). Caldwell Pa., *Shinners 22587* (SMU). Cameron Pa., *Thieret 8935* (RSA, SMU). Catahoulla Pa., 11 May 1912, *Meeker s.n.* (LSU). Claiborne Pa., *Arant 87* (LSU). DeSoto Pa., *Shinners 22847* (SMU). East Baton Rouge Pa., *Montz 1035* (LSU, SMU). East Carroll Pa., *Lonard 8783* (TAES). Evangeline Pa., *Iltis 21646* (WIS). Franklin Pa., *Watters 1123* (LSU). Grant Pa., *Bonner 204* (LSU). Iberia Pa., *Brown 1840* (MICH). Iberville Pa., *Omar 32* (LSU). Jackson Pa., *Moore 12* (MO). Jefferson Pa., *Ball 342* (CAN, F, GH, MIN, MO, NY, US). Jefferson Davis Pa., *Palmer 7648* (CAS, MIN). Lafayette Pa., *Thieret 10286* (DS, FSU, SMU). LaSalle Pa., *Shinners 27187* (SMU). Lincoln Pa., *Hardin 492* (GH, MICH, NCU, TEX, US). Livingston Pa., *Thieret 32952* (A, DUKE, FSU, TEX). Morehouse Pa., *Demaree 20964* (MO, NY, POM, SMU). Natchitoches Pa., *Palmer 7302* (CAS, MO, NY). Orleans Pa., *Demaree 34657* (FSU, GH, OKLA, RSA, SMU, USF, VDB). Ouachita Pa., *Kral 8428* (GH, FSU, SMU, VPI). Pointe Coupee Pa., *Chaney 139* (LSU). Rapides Pa., *Jones 193* (ILL). Red River Pa., *Shinners 28279* (SMU). Sabine Pa., *Shinners 22802* (SMU). St. Charles Pa., *Montz 822* (LSU). St. Helena Pa., *Allen 691* (LSU). St. John the Baptist Pa., *Gugich 255* (LSU). St. Mary Pa., *Dooley 12* (NCU). St. Tammany Pa., *Thieret 29169* (DS, GA). Tangipahoa Pa., *Wilson 1066* (FSU). Terrebonne Pa., *Holmes 109* (LSU). Union Pa., *Lester 70* (UARK). Vernon Pa., *Cooley & Brass 3946* (USF). Washington Pa., *Shinners 29214* (SMU). Webster Pa., *Shinners 26747* (SMU). West Baton Rouge Pa., *Allison 313* (MISSA). West Carroll Pa., 5 Jul 1910, *Moseley s.n.* (F). West Feliciana Pa., *McNiel 33* (GH). Winn Pa., *Shinners 20281* (SMU).—Maine: Cumberland Co., *Chamberlain 417* (GH). Hancock Co., 14 Sep 1896, *Rand s.n.* (UC, YU).—Maryland: Anne Arundel Co., *Blake 11358* (CAS, NA, POM, TEX, TTC). Baltimore Co., 30 May 1890, *Metcalf s.n.* (MU). Baltimore City, 30 May 1889, *Metcalf s.n.* (MU). Calvert Co., *Seymour 16776* (CAS, MIL, MO, WIS). Cecil Co., *Popowsky 25* (PH). Charles Co., *Hill 2582* (DAO). Dorchester Co., *Milby 18* (OKL). Kent Co., 8 Aug 1902, *Vanatta s.n.* (PH). Prince George Co., *Pollard 810* (MSC, NY). St. Mary's Co., *Archer 7494* (NA). Somerset Co., 18 Jul 1890, *Holmes s.n.* (MSC). Talbot Co., *Earle 3546* (GH, PH). Wicomico Co., 13 Jun 1906, *Carter s.n.* (PH, SMU). Worcester Co., *Perdue 664* (NA).—Massachusetts: Dukes Co., *Bicknell 6620* (NY). Hampden Co., 30 Jun 1914, *Andrews s.n.* (POM). Middlesex Co., 13 Jun 1911, *Fletcher s.n.* (GH). Norfolk Co., *Blake 1700* (TEX). Suffolk Co., 1878, *Faxon & Faxon s.n.* (GH). Worcester Co., 15 Sep 1926, *Woodward s.n.* (GH).—Michigan: Barry Co., *Drew M16* (MSC). Berrien Co., *Swink 64* (MICH). Cass Co., *Rapp 395* (MICH). Kalamazoo Co., *Hott 36* (MSC). Kent Co., *Bazuin 438* (MSC). Leelanau Co., 11 Aug 1953, *Hanes s.n.* (MICH). Lenawee Co., *Horne & Horne 172* (MICH). Monroe Co., 15 May 1958, *Churchill s.n.* (MSC). Montcalm Co., *Bazuin 3097* (F). Newago Co., *Hermann 8617* (MICH, NY). Oakland Co., 14 Aug 1901, *Cooper s.n.* (MIN). Oceana Co., *Yuncker 741* (ILL). St. Clair Co., 20 Jun 1920, *Billington s.n.* (MICH). Washtenaw Co., *Hoseney 1* (MICH).—Minnesota: Anoka Co., *Sharsmith 5402* (MIN).—Mississippi: Adams Co., *Jones & Jones 4243* (MISS). Alcorn Co., *Pullen 64476* (MISS). Amite Co., *Jones & Jones 4218* (MISS). Attala Co., *Jones 16999* (MISS). Carroll Co., *Clute 33* (C, CAS, F, NY). Chocktaw Co., 19 May 1904, *Jensen s.n.* (MO). Clarke Co., *Jones & Jones 4712* (MISS). Coahoma Co., *Pullen 70460* (MISS). Copiah Co., *Jones 4433* (MISS). DeSoto Co., *Pullen 70285* (MISS). Forrest Co., *Hudson 60* (FSU). Franklin Co., *Temple 8863* (MISS). George Co., *Jones & Jones 11284* (MISS, VDB). Greene Co., *Jones 5592* (MISS). Grenada Co., 22 Apr 1932, *Millsaps s.n.* (NCU). Hancock Co., *Allison 3121* (MISS). Harrison Co., *Demaree 29022* (ALU, MIN, OKL, OKLA, RSA, SMU, TEX). Hinds Co., *Temple 8109* (MISS). Holmes Co., *Pullen 64737* (AUA, MISS, NCU). Humphreys Co., *Pullen 69369* (MISS). Issaquena Co., *Pullen 70152* (MISS). Jackson Co., *Skehan 83* (CAS, CU, DUKE, F, GH, IA,

ILL, KANU, MIL, MO, MSC, NCU, SMU, TEX, UMO, WIS). Jasper Co., *Jones 5936* (MISS). Jefferson Co., *Noble et al. 11561* (MISS). Jefferson Davis Co., *Jones & Reynolds 11723* (GA, MISS). Jones Co., *Jones 4469* (MISS). Kemper Co., *Jones 15995* (MISS, VDB). Lafayette Co., *Schuster A-7608* (DUKE, VDB). Lamar Co., *Jones 2705* (MISS). Lauderdale Co., *Hood 4669* (FLAS). Lawrence Co., *Rhodes 12323* (LSU). Leake Co., *Jones et al. 16879* (MISS). Leflore Co., *Pullen 6839* (MISS). Lincoln Co., *Flint s.n.* (MASS). Marion Co., *Jones & Reynolds 11142* (MISS). Marshall Co., *Lassetter 210* (MISS). Monroe Co., *Temple 4782* (MASS, MISS, NCU). Neshoba Co., *Jones 16269* (MISS). Newton Co., *Jones 18393* (GA, MISS). Noxubee Co., 26 Apr 1962, *Ugent & Luther s.n.* (WIS). Oktibbeha Co., *Tracy 131* (ARIZ, CS, CU, F, GH, ILL, KSC, LSU, MICH, MSC, NY, PAC, PH, VPI, VT, WIS, WVA). Panola Co., *Pullen 7082* (MISS). Pearl River Co., *Demaree 50050* (DS, KY, SMS, SMU). Pike Co., *Wagenknecht 3553* (KANU). Quitman Co., *Pullen 67104* (MISS). Rankin Co., *Jones 18625* (MISS, VDB). Sharkey Co., *Pullen 70115* (MISS). Simpson Co., *Thomas et al. 7021* (CM, DS). Smith Co., *Jones 4042* (MISS). Stone Co., *Jones & Jones 11161* (MISS). Tallahatchie Co., *Pullen 6868* (MISS). Tate Co., *Temple 2935* (MISS). Tishomingo Co., *Coleman 46328* (TENN). Tunica Co., *Temple 5047* (MISS). Union Co., *Pennell 99* (MISS). Washington Co., *Gunn 10537* (MISS). Wayne Co., *Jones 4918* (MISS). Webster Co., 10 May 1928, *Finster s.n.* (MICH).—MISSOURI: Barry Co., *Palmer, E. J. 62155* (UMO). Barton Co., *Palmer, E. J. 65744* (UMO). Bates Co., *Bush 15503* (MO). Bollinger Co., 30 May 1898, *Russell s.n.* (MO). Boone Co., 14 May 1927, *Rickett s.n.* (UMO, WIS). Buchanan Co., *Henderson 65–396* (KANU, SMS, UMO). Butler Co., *Smith 523* (F). Camden Co., *Stephens 56429* (KANU). Cape Girardeau Co., *Drushel 3191* (ILL, MO). Carroll Co., *Henderson 66–306* (FSU, G, KANU, SMS, VT). Carter Co., *Redfearn 5732* (FSU, SMU). Cedar Co., *Redfearn & Houk 8166* (FSU, SMS, UMO, VDB). Chariton Co., 2 Jul 1925, *Young s.n.* (CM). Clay Co., *MacKenzie 288* (MO, NY). Dade Co., *Palmer, E. J. 55695* (UMO). Dent Co., *Maupin 578* (SMS, UMO). Dunklin Co., 12 Sep 1893, *Bush s.n.* (MO). Franklin Co., 2 Jun 1927, *Kellogg s.n.* (MO). Gasconade Co., *Chandler 4447* (ASTC). Greene Co., *Redfearn 3591* (FSU, SMS). Grundy Co., *Crookshanks 4* (MO). Iron Co., *Smith 412* (F). Jackson Co., *Bush 7550* (GH, ILL, MIN, NY, US). Jasper Co., *Palmer 301* (MIN, MO). Jefferson Co., *Ownbey 743* (MIN). Johnson Co., *Campbell 108-2* (MO). Laclede Co., 21 Jun 1938, *Moore s.n.* (F). Lawrence Co., *Semple 539* (DUKE, MO). Lincoln Co., *Davis 1397* (MO). Livingston Co., *Sparling 838* (F). Madison Co., 19 May 1927, *Beardsley & Larsen s.n.* (G, MIN, MO). Marion Co., *Davis 4364* (DS, MIN, MO). McDonald Co., *Kellogg 25889* (MO, UMO). Newton Co., *Wilkins 53* (TTC). Oregon Co., *Thomas 9942* (TENN). Ozark Co., *Bush 13376* (MO, TEX). Pemiscot Co., *Demaree 30788* (KANU, MICH, MSC, NCU, OKL, RSA, SMU, TENN, TEX, UMO, WVA). Perry Co., Mar 1886, *Demetrio s.n.* (F). Phelps Co., *Muenscher & Winne 16639* (CU). Pike Co., 1948, *Etter s.n.* (MO). Platte Co., *Henderson 65–133* (FSU, ISC, KANU, SMS, UMO). Polk Co., *Henderson 67-348* (CAS, SMS, UMO, VT). Reynolds Co., *Smith 268* (F). St. Charles Co., *Dowling 1958* (UMO). St. Clair Co., 22 May 1932, *Drouet s.n.* (UMO). St. Genevieve Co., *Brinker 1660* (ILLS). St. Louis City, *Mühlenbach 3316* (MO). St. Louis Co., 28 May 1887, *Eggert s.n.* (MIN, NCU). Scott Co., *Rhodes 12409* (LSU). Shelby Co., *Palmer, E. J., & Steyermark 40911* (MO). Stone Co., *Bush 15630* (MO). Taney Co., 23 Aug 1940, *Spencer s.n.* (UMO). Vernon Co., 26 May 1871, *s.c., s.n.* (DS, MO).—NEBRASKA: Adams Co., *Stephens 47389* (KANU). Antelope Co., 8 Jun 1929, *Wernecke s.n.* (MIL). Brown Co., 27 Aug 1908, *Bates s.n.* (GH). Cass Co., *Stephens & Brooks 32143* (CAS, KANU). Cherry Co., Jun 1890, *Bates s.n.* (E). Colfax Co., 10 Jun 1956, *West s.n.* (CU). Cuming Co., *Stephens & Brooks 38613* (KANU). Custer Co., 17 Jun 1901, *Bates s.n.* (GH). Devel Co., *Rydberg 109* (US). Dodge Co., *Kiener 14702* (WIS). Douglas Co., 13 Jul 1947, *Kirk s.n.* (WIS). Franklin Co., Jun 1928, *Hapeman s.n.* (PH, VDB). Hall Co., *Lemaire 2319* (KSC). Jefferson Co., *Stephens 53822* (KANU). Kearney Co., 22 Jun 1932, *Hapeman s.n.* (ARIZ, ND, TENN, UC). Keya Paha Co., *Clements 2861* (CU, GH, ISC, MIN, NY). Lancaster Co., 10 Jun 1898, *Sheldon s.n.* (WVA). Lincoln Co., 25 May 1903, *Mill s.n.* (US). Madison Co., *Harms 587* (KANU, NY). Nance Co., *Osborn 1163R* (MO). Nuckolls Co., *Stephens 65294* (KANU). Perkins Co., *Stephens 79036* (KANU). Red Willow Co., *Stephens 65395* (KANU). Richardson Co., *Shildneck C-6902* (ILLS, KANU). Sarpy Co., *Stephens 20900* (KANU). Thomas Co., 16 Jun 1904 *Mell s.n.* (ILL). Webster Co., *McGregor 19331* (KANU).—NEW JERSEY: Atlantic Co., *Gershoy 513* (CU, GH). Burlington Co., *Dreisbach 981* (MICH, PH). Camden Co., 22 May 1922, *Bassett s.n.* (CM, GH). Cape May Co., 1 Jun 1920, *Macfarlane s.n.* (PENN). Cumberland Co., *Adams 1000* (GH, PH). Gloucester Co., *Fogg 8577* (PENN, PH). Hunterdon Co., *Long 39994* (DS, PH). Mercer Co., 13 Jun 1903, *Logan s.n.* (WVA). Middlesex Co., 4 Jul 1931, *Loughridge s.n.* (MU). Monmouth Co., *Dreisbach 952* (CM, F). Ocean Co., 9 Jun 1970, *Laport s.n.* (NCU). Salem Co., *Long 49974* (PH).—NEW YORK: Bronx Co., N.Y. Botanical Garden, *Gilly 324* (NY). Chemung Co., Elmira College, *Munz 1946* (CU). Erie Co., Buffalo,

Muenscher 17257 (CU). Nassau Co., Valley Stream, *Bicknell 6619* (NY, PH). Suffolk Co., Sag Harbor, *Ferguson 6788* (NY).—NORTH CAROLINA: Alamance Co., Swepsonville, *Radford 10683* (NCU). Alexander Co., Taylorsville, *Keever 309* (DUKE). Anson Co., Lilesville, *Radford 13503* (NCU). Beaufort Co., Blount's Creek, *Godfrey et al. 7040* (DUKE, F, GH, NY, POM, UC, US). Bertie Co., Merry Hill, *Correll 1985* (DUKE). Bladen Co., Elizabethtown, *Ahles & Ramseur 23515* (NCU). Brunswick Co., Shallotte, 1954, *Henry s.n.* (CM). Buncombe Co., Biltmore, 1896, *s.c., s.n. 672* (CU, F, MIN, US, VT). Burke Co., Morganton, *Bell 6428* (NCU). Cabarrus Co., Kannapolis, *Bell 2260* (NCU). Caldwell Co., Oak Hill, *Radford 14907* (NCU). Carteret Co., Morehead City, *Crutchfield 565* (NCU). Caswell Co., Estelle, *Ahles 54057* (NCU). Catawba Co., Conover, *Bell 6715* (NCU). Chatham Co., Farrington, *Ahles & Haesloop 53335* (ALU, AUA, CM, DAO, FLAS, FSU, GA, GH, IA, ILL, MASS, NY, PAC, SMU, US, USF, VDB). Cherokee Co., Murphy, *Ahles & Radford 13035* (NCU). Chowan Co., Edenton, *Ahles & Ashworth 39675* (NCU). Cleveland Co., Lawndale, *Ahles & Leisner 15192* (NCU). Columbus Co., Fairbluff, *Wilbur 5283* (DUKE). Cumberland Co., Fayetteville, *Wilbur 5604* (DUKE). Currituck Co., Knotts Island, *Correll 2145* (TENN). Davidson Co., Silver Valley, *Radford 12769* (NCU). Davie Co., S of fork on Yadkin River, *Radford 10869* (NCU). Duplin Co., Warsaw, *Martin 299* (OKL). Durham Co., Braggtown, *Radford 44784* (CM, DAO, FLAS, FSU, GA, KE, NY, SMU, US, USF, VDB). Edgecombe Co., Rocky Mount, *Radford 33918* (NCU). Forsyth Co., Kernersville, *Ahles & Britt 40642* (NCU). Franklin Co., New Hope, 1956, *Mueller s.n.* (WIS). Gaston Co., Bessemer City, *Ahles & Leisner 15113* (NCU). Gates Co., Sunbury, *Ahles & Ashworth 40197* (NCU). Granville Co., Creedmoor, *Ahles & Radford 11437* (NCU, UARK). Greene Co., Snow Hill, *Correll 1367* (DUKE). Guilford Co., Jamestown, *Bell 11738* (NCU, VPI). Halifax Co., Scotland Neck, *Correll 2391* (BHO, GH). Hamett Co., Lillington, *Laing 988* (TEX). Henderson Co., Hendersonville, *Pittillo 117* (NY, VDB). Hertford Co., Murfreesboro, *Correll 2317* (DUKE). Hoke Co., Ashley Heights, *Ahles & Neuber 24962* (NCU). Hyde Co., Scranton, *Radford 33651* (NCU). Iredell Co., Statesville, *Ahles & Britt 40928* (NCU). Johnston Co., Smithfield, *Wilbur 5488* (DUKE). Jones Co., Tuckahoe Swamp, *Radford 36975* (NCU). Lee Co., Broadway, *Stewart 160* (NCU). Lenoir Co., La Grange, *Radford 22103* (NCU). Lincoln Co., Flay, *Bell 6591* (NCU). Madison Co., Hot Springs, 1899, *Churchill s.n.* (TENN). Martin Co., Williamston, *Radford 32263* (NCU). McDowell Co., Ashford, *Hunnewell 11197* (GH). Mecklenburg Co., Charlotte, *Ahles & Britt 38726* (CM, NCU). Montgomery Co., Uwharrie, *Wells 2849* (NCU). Moore Co., Whispering Pines, *Carter III 485* (NCU). Nash Co., Little Peachtree Creek, *Ahles & Horton 11741* (NCU). New Hanover Co., Wilmington, *Bell 12893* (MIN, NCU). Northampton Co., 1.4 mi W of jct. U.S. 158 & 258, *Ahles & Duke 41733* (NCU). Onslow Co., Jacksonville, *Ahles & Ramseur 24098* (NCU). Orange Co., Chapel Hill, *Radford & Stewart 339* (NCU, PAC). Pamlico Co., Janeiro, *Radford 31964* (NCU). Pasquotank Co., Elizabeth City, *Ahles & Ashworth 39998* (NCU). Pender Co., Maple Hill, *Wilbur 5333* (DUKE). Perquimans Co., 1932, *Glasson s.n.* (DUKE). Pitt Co., Greenville, *Lamm 11* (NCU). Polk Co., Tryon, *Walker 3420* (US). Richmond Co., Hoffman, *Radford 11329* (NCU). Robeson Co., St. Pauls, *Britt 254* (E). Rockingham Co., Reidsville, *Radford 13620* (C). Rowan Co., Spencer, *Ahles & Radford 12910* (NCU). Rutherford Co., Union Mills, 1947, *Woodbury & Buswell s.n.* (CONN). Sampson Co., 12 mi N of Salemburg, *Wilbur 5232* (DUKE). Scotland Co., Silver Hill, *Ahles & Hammond 24720* (NCU). Stanly Co., Albemarle, *Radford 10571* (NCU). Stokes Co., Walnut Cove, *Radford 33015* (NCU). Swain Co., Mica Knob Road, *Ahles & Bell 14159* (NCU). Tyrrell Co., Columbia, *Radford 33760* (NCU). Union Co., Unionville, *Ahles & Radford 11959* (NCU). Vance Co., 2 mi SE of Fairport, *Ahles & Bell 12713* (NCU). Wake Co., Raleigh, *Buell 1350* (PH). Washington Co., Plymouth, *Radford 32335* (NCU). Wayne Co., Mt. Olive, *Barwick 84* (MISS). Wilson Co., Wilson, *Reed 100083* (NCU).—NORTH DAKOTA: Billings Co., Medora, *Godfrey 5626* (NDA). Morton Co., *Bell 458* (POM).—OHIO: Adams Co., Sandy Springs Cemetery, *Cusick 12991* (KE). Athens Co., Athens, *Hall 1222* (BHO). Belmont Co., Barnesville, *Laughlin 1344* (OS). Butler Co., flood plain of Indian Creek, *Cobbe & Cobbe 83* (B, C, DS, MU, UC). Coshocton Co., 1930, *Shelby s.n.* (OS). Cuyahoga Co., Berea, 1895, *Claassen s.n.* (OS). Fairfield Co., Pleasant Run Bog, Berne Twp., 1955, *Goslin s.n.* (BHO). Gallia Co., Perry Twp, *Silberhorn 3771* (KE). Hamilton Co., Camp Dennison, *Braun 1344* (OS). Henry Co., Hoy Cemetery, Harrison Twp, *Cusick 11850* (KE, MU). Highland Co., Hillsboro, 1926, *Roads s.n.* (OS). Jackson Co., Liberty Twp, *Bartley 1508* (OS). Lake Co., Perry, 1945, *Tyler s.n.* (OS). Lucas Co., Whitehouse, 1925, *Moseley s.n.* (CM, MICH, US). Pickaway Co., Circleville, 1960, *Bartley s.n.* (OS). Ross Co., Paxton Twp, 1932, *Bartley s.n.* (BHO). Trumbull Co., Brocerville, *Rood 649* (KE, MIN). Wood Co., Bowling Green, 1921, *Moseley s.n.* (GH, MICH).—OKLAHOMA: Adair Co., Westville, *Wallis 8129* (KANU, NCU, OKL, OKLA, TEX, VDB). Alfalfa Co., Jet, *Stephens 21591* (DS, KANU). Atoka Co., Atoka, *Cinq-Mars 73–16* (CAN). Beaver Co., Slapout, *Stephens 74810*

(KANU, OKL). Beckham Co., Sayre, *Hopkins & Van Valkenburgh 5783* (NY, OKL). Blaine Co., Greenfield, *Engleman 1405* (OKLA). Bryan Co., Durant, *Blain 115* (POM). Caddo Co., rim of Devil's Canyon, *Hopkins et al. 274* (DS, F, MO, OKLA, SMU, TEX, UC, US, WIS). Canadian Co., Hinton, *Hopkins 1421* (OKL). Cherokee Co., Fort Gibson, *Wallis 6643* (KANU, NCU, OKL, SMU, TEX, UARK). Choctaw Co., Hugo, 1950, *Gates s.n.* (OKL). Cleveland Co., Norman, 1924, *Bruner s.n.* (ISC, OKL, OKLA, US). Coal Co., Lehigh, *Hopkins et al. 1067* (DAO, OKL, SMU, TEX, WIS). Comanche Co., Fort Sill, *Clemens 11701* (MO). Cotton Co., Randlett, *Shinners 25976* (SMU). Creek Co., Sapulpa, *Waterfall 2056* (NY, OKL, OKLA). Custer Co., Weatherford, *Waterfall 2958* (OKL, OKLA). Delaware Co., Grove, *Wallis 2916* (OKLA). Ellis Co., Shattuck, *Clifton 3103* (MIN). Garfield Co., Covington, *Stephens 74873* (KANU, OKL). Garvin Co., Stratford, *Duffer 373* (OKLA). Grady Co., Tuttle, *Pearce 685* (KSC, OKL). Grant Co., Medford, *Stephens 74828* (KANU, OKL). Greer Co., Willow, *Hixson 37* (OKLA). Harmon Co., McQueen, *Stephens 20779* (DS, KANU). Harper Co., Laverne, *Isaac 93* (OKLA). Haskell Co., Kinta, *Barkley 1265* (OKL). Hughes Co., Wetumka, *Wiedeman 367* (OKL, OKLA). Jackson Co., Headrick, *Hixson 68* (OKLA). Johnston Co., Tishomingo, *Robbins 2398* (NY, OKL, SMU, TAES, UC). Kay Co., Blackwell, *George 2* (OKL). Kingfisher Co., Huntsville, *Blankinship 17298* (GH, MO, MU, US, WIS). Kiowa Co., Snyder, *Hopkins 3024* (GH, OKL). Latimer Co., Red Oak, *Means, Jr. 3395* (OKLA). Le Flore Co., Page, *Stevens 1392* (DS, GH, ILL, MIN, MO, OKL, OKLA). Lincoln Co., Perkins, *Payton 86* (OKLA). Logan Co., Guthrie, *Beck 89* (OKLA). Love Co., Marietta, *Stevens 95* (GH). Marshall Co., Lake Texoma, *Goodman 5793* (ILL, MIN, OKL). Mayes Co., Peggs, *Wallis 6694* (OKLA). McClain Co., Purcell, *Hopkins 44* (OKL). McCurtain Co., Broken Bow, *Dreher 348* (LTU). McIntosh Co., Checotah, *Crook & Crook 1068* (OKL). Murray Co., Davis, *Randel 49* (OKL). Muskogee Co., Fort Gibson, *Wallis 7449* (FSU, KANU, NCU, OKL, OKLA, SMU, VDB). Oklahoma Co., Oklahoma City, 1892, *Shimek s.n.* (F, ISC, MO. US). Osage Co., Tulsa, *Perino & Perino 416* (KANU, OKL). Ottawa Co., Grove, *Wallis 6961–1* (OKLA). Payne Co., Stillwater, 1896, *Bogue s.n.* (MIN). Pittsburg Co., 1935, *McClary s.n.* (OKL). Pontotoc Co., Ada, *McCoy 2624* (OKLA). Pottawatomie Co., Tecumseh, *Barkley 52* (MO, OKL). Pushmataha Co., Clayton, *Means, Jr. 3339* (OKLA). Rogers Co., Claremore, *Fogg, Jr. 18773* (PENN). Sequoyah Co., Gore, *Willis 8128* (GH, KANU, NCU, OKL, OKLA, SMU, VDB). Stephens Co., Marlow, *Rice 39* (OKL). Tillman Co., Chattanooga, 1969, *Crook & Crook s.n.* (OKL, SMU). Tulsa Co., Jenks, *Clark 278* (OKL, OKLA). Wagoner Co., Wagoner, *Bebb 3948* (OKL). Washington Co., Bartlesville, *McDonald 271A* (OKLA). Washita Co., *Eskew 1618* (OKL, SMU). Woods Co., Alva, *Stevens 618* (DS, ILL, MIN, MSC, OKL, OKLA, SMU, US). Woodward Co., Sharon, *Stephens 74739* (KANU, OKL).—PENNSYLVANIA: Allegheny Co., Pittsburgh, 1919, *Patterson s.n.* (CM). Berks Co., Maidencreek Station, *Wilkens 8204* (PENN, PH). Bucks Co., Bristol, 1898, *Fretz s.n.* (PENN, PH). Chester Co., Brookfield, 1930, *Stone s.n.* (GH, PENN, PH). Delaware Co., Swarthmore, 1905, *Cresson, Jr. s.n.* (PH). Lancaster Co., Safe Harbor, 1960, *Wherry s.n.* (PENN). Monroe Co., Shawnee on Delaware, 1942, *Dimmick s.n.* (SMU). Montgomery Co., Ambler, *Long 59458* (PH). Northampton Co., Riverton, *Schaeffer 45374* (PH). Philadelphia Co., Somerton, *Long 31396* (PH).—SOUTH CAROLINA: Abbeville Co., *Chitterden 17* (NCU). Aiken Co., Aiken, *Brown 2* (DAO). Allendale Co., Fairfax, *Ahles & Bell 10618* (NCU). Anderson Co., Anderson, 1920, *Davis s.n.* (DS, ND, POM). Bamberg Co., Bamberg, *Ahles & Haesloop 22116* (NCU). Barnwell Co., Williston, *Radford 9193* (NCU). Beaufort Co., Hunting Island, *Rossbach & Murphy 2384* (NCU, WVA). Berkeley Co., SW corner of county, *Ahles & Haesloop 22342* (NCU). Calhoun Co., St. Matthews, *Ahles & Haesloop 21790* (NCU). Charleston Co., Charleston, *Robinson 194* (GH). Cherokee Co., Blacksburg, 1917, *Munz s.n.* (CU). Chester Co., Leeds, *Bell 7315* (NCU). Chesterfield Co., Cheraw, *Duke 565* (KY). Clarendon Co., Manning, *Stone 497* (PH). Colleton Co., Ritter, *Bell 1809* (NCU). Darlington Co., Darlington, 1897, *Ward s.n.* (US). Dillon Co., Hamer, *Ahles & Ramseur 23161* (NCU). Dorchester Co., Middleton, *Duncan 6025B* (GA). Edgefield Co., Edgefield, *Radford 20342* (NCU). Fairfield Co., Strother, *Bell 7095* (NCU). Florence Co., Johnsonville, *Ahles 40432* (NCU, SMU). Georgetown Co., Murrells Inlet, *Weatherby & Griscan 16594* (GH, USF). Greenville Co., Greenville, *Rodgers 67030* (MISS, NCU). Hampton Co., Garnett, *Ahles et al. 57883* (NCU). Horry Co., Socastee, *Bell 6164* (NCU). Kershaw Co., 10 mi E of Camden, *Radford 20756* (NCU). Lancaster Co., Pageland, *Wilbur 8942* (DUKE). Laurens Co., Clinton, *Bell 7929* (NCU). Lexington Co., Batesburg, *McGregor 21* (US). Marion Co., Little Pee Dee River at Potato Bed Ferry Bridge, *Bell 6288* (NCU). Marlboro Co., Gibson, *Radford 12697* (NCU). Newberry Co., Newberry, *Bell 6876* (NCU). Oconee Co., Newry, *House 2208* (NY, US). Orangeburg Co., Orangeburg, *Ahles & Haesloop 21504* (NCU). Pickens Co., 2.4 mi S of N.C.-S.C. state line on U.S. Rte 178, *Ahles & Bell 14306* (NCU). Richland Co., Columbia, *Boufford et al. 12775* (NCU). Saluda Co., Saluda, *Radford*

20645 (NCU). Spartanburg Co., Spartanburg, *Faust 110* (DUKE). Sumter Co., Rembert, *Radford 20906* (NCU). Union Co., Carlisle, *Bell 8392* (IND). Williamsburg Co., Greeleyville, *Radford 21299* (UC). York Co., Bethany, *Ahles & Haesloop 22823* (NCU).—South Dakota: Custer Co., Reeves Canyon, *Barr 73* (BH). Fall River Co., Hot Springs, *Rydberg 703* (US). Harding Co., Redig, *Moore 2076* (MIN). Meade Co., Faith, *Moyer 229* (MIN). Pennington Co., Rapid City, *Moore 713* (MIN).—Tennessee: Anderson Co., Melton Hill Reservoir, *Ellis 28862* (TENN). Bedford Co., Shelbyville, *Kral 52542* (VDB). Benton Co., Holladay, *Sharp et al. 12881* (TENN). Blount Co., Maryville, 1934, *Godfrey s.n.* (TENN). Campbell Co., Morley, 1923, *Bright s.n.* (CM). Cheatham Co., Big Marrowbone Creek Road Bridge, *Mulcahy 131* (NCU, VDB, VPI). Cocke Co., Wolf Creek, *Wilson 1431* (TENN). Coffee Co., Tullahoma, *Sharp et al. 3738* (TENN). Davidson Co., Belleview, *Demaree 47478* (DAO, DS, MASS, MIN, UMO, VDB). Dyer Co., Bogota, *Sharp et al. 12196* (TENN). Fayette Co., Jopic School, Ames Plantation, *Hebb 325* (TENN). Franklin Co., Highland Rim Forest Agricultural Exp. Station, *Ratledge & DeSelm 30754* (TENN). Gibson Co., Trenton, *Sharp et al. 12496* (TENN). Giles Co., Prospect, *Quarterman 5260* (VDB). Greene Co., Greenville, *Davis & Mahler 4426* (MSC, NCU, SMU). Hamilton Co., Chattanooga, *Williams 1398B* (GH). Hawkins Co., Stanley Valley, *Wolfe 19103* (TENN). Henry Co., Elkhorn, *Sharp et al. 13007* (TENN). Hickman Co., Bucksnort, *Kral 45642* (VDB). Knox Co., Knoxville, 1893, *Ruth s.n.* (MICH, MIN). Lauderdale Co., *Sharp et al. 12133* (TENN). Lewis Co., Meriwether Lewis Nat. Monument, *King 88* (VDB). Madison Co., U.T. Exp. Station, *Sharp et al. 12614* (TENN). Marion Co., Monteagle, 1931, *Jennison & Sharp s.n.* (WVA). McMinn Co., Athens, *Sharp & Hesler 1012* (MO, TENN, UC). Meigs Co., Decatur, *Sharp & Jones 28221* (TENN). Montgomery Co., Clarksville, *Chester 2020* (NCU, SMU, TENN). Morgan Co., Oliver Springs, *Diggs 25* (UC). Obion Co., Walnut Log, *Eyles & Eyles 8369* (BHO, GH). Rhea Co., Watts Bar Dam, *Shanks et al. 4199* (TENN). Roane Co., Clinch River Breeder Reactor Site, *Hale 47619* (TENN). Rutherford Co., Eagleville, *Demaree 45783* (NCU, SMU, TENN, VDB). Sevier Co., Sugarloaf Mountain, 1964, *Thomas s.n.* (SMU, TENN). Shelby Co., Memphis, *Demaree 21374* (ARIZ, DUKE, ISC, LSU, MIN, MO, ND, NY, OKL, POM, UARK, WIS). Sumner Co., Mitchellville, *Shanks et al. 14303* (RSA, TENN). Tipton Co., Richardsons, *Sharp et al. 12166* (TENN). Trousdale Co., 12 mi S of Macon Co. Line on 141, *Blum 3470* (PH, VDB). Unicoi Co., Erwin, 1931, *Wiegand s.n.* (CU). Williamson Co., 20 mi W of Nashville, *Anrett 63* (VDB).—Texas: Anderson Co., Palestine, 1935, *Smith s.n.* (DS, OKLA, TTC). Angelina Co., Zavalla, *Wilson 73* (TAES). Aransas Co., Rockport, *Raven & Gregory 19395* (DS, RSA, US). Atascosa Co., Pleasanton, *Schulz 308* (US). Austin Co., S.A. Austin Park, *Sinclair et al. 1527* (UMO). Bastrop Co., Butler, *McCart 6378* (SMU). Bee Co., Skidmore, 1905, *Lewton s.n.* (US). Bell Co., Temple, *Wolff 836* (TAES, US). Bexar Co., San Antonio, 1932, *Clare s.n.* (WVA). Bosque Co., Walnut Springs, *Van Vleet 51* (SMU). Bowie Co., DeKalk, 1960, *Mehone s.n.* (ASTC). Brazoria Co., Columbia, *Palmer 5016* (CS, CU, MIN, POM). Brazos Co., College Station, *Massey 500* (NCU, OKL, TAES). Brewster Co., Alpine, *Sperry T603* (GH, TAES, UC). Briscoe Co., Quitaque, *Whitehouse 10009* (SMU). Burnet Co., Burnet, *Lathrop 2019* (KANU). Caldwell Co., Lockhart, *McCart 6455* (SMU). Calhoun Co., Long Mott, *Gentry 950* (ARIZ). Callahan Co., Clyde, *Henderson 62–191* (FSU). Cameron Co., Virginia Point, *Bray 2* (US). Camp Co., Pittsburg, *Shinners 13983* (SMU). Cass Co., Atlanta, *Demaree 53841* (DS, SMU). Cherokee Co., Sacul, 1958, *Lyles s.n.* (DS). Childress Co., Memphis, *Higgins 7251* (NY). Clay Co., Byers, *Stephens 20494* (DS, KANU). Colorado Co., Columbus, *Shinners 14637* (SMU). Comal Co., New Braunfels, *Lindheimer 810* (MIN, MO). Comanche Co., De Leon, *Stanford 2234* (OKLA). Cottle Co., *Rowell, Jr. 8003* (OKLA, TEX, TTC). Dallam Co., Texline, *York & Rodgers 202* (TTC). Dallas Co., Seagoville Road, *Lundell & Lundell 8356* (DS, FSU, GH, MICH, MIL, NY, POM, TEX, UC). Denton Co., Lewisville, *Whitehouse 19894* (GA, OKLA, SMU). De Witt Co., Cuero, *Howell 328* (US). Dimmit Co., Carrizo Springs, *Palmer 33731* (NY). Duval Co., Falfurrias, *Cory 55282* (SMU). Edwards Co., Rocksprings, *Cory 3277* (GH). Erath Co., Stephenville, *Mertins 1008* (OKLA). Fannin Co., Monkstown, *Rochat & Sinclair 50* (TEX). Fayette Co., La Grange, *Rhodes 14629* (LTU). Fort Bend Co., Arcola, *Palmer 5097,* pro parte (MO, POM). Frio Co., Dilley, *Lucas et al. 14203* (RSA). Galveston Co., Texas City, *Turner 1793* (SMU). Gillespie Co., Willow City, *Nixon G12* (TEX). Goliad Co., Goliad, *Williams 111* (PH, TEX). Gonzales Co., Ottire, *Warnock 20548* (TEX). Grayson Co., Denison, *Gentry 141* (SMU, TEX). Gregg Co., Longview, 1899, *Eggert s.n.* (MO). Grimes Co., Navasota, 1934, *Flower contest* (MICH, MIN, MO). Hardin Co., Votaw, *Pratt 76* (TAES). Harris Co., Houston, *Traverse 1396* (GH, SMU, TEX). Harrison Co., Marshall, *Murtishaw 185* (LTU). Hartley Co., Dalhart, *Jones 3A* (GH). Haskell Co., Rule, *Cory 37210* (POM, TAES). Hemphill Co., Gene Howe Wildlife Management Area, *Rowell, Jr. 5376A* (OKLA, TTC). Henderson Co., La Rue, *Sanders 140* (MICH, SMU). Hidalgo Co., Samfordyce, *Small*

& Wherry 11918 (NY, TEX). Hood Co., Granbury, *Shinners 11123* (SMU). Houston Co., *Parks 124* (TAES). Hunt Co., Greenville, *Bebb 2651* (OKL, WIS). Jasper Co., Kountze, *Demaree 55414* (DS, MASS). Jeff Davis Co., Madera Springs, *Cory 18440* (POM). Jefferson Co., Cheek, *Lundell & Lundell 10924* (MICH, POM, SMU, TEX). Karnes Co., 12 mi S of Kenedy, *Cory 54085* (KANU, GA, IND, OKLA, SMU, TEX). Kaufman Co., Kemp, *Shinners 10872* (SMU). Kleberg Co., Kingsville, *Sinclair O-3* (TEX). Lee Co., Farm Road 619, *McCart 6416* (SMU). Leon Co., Jewett, *Koelling 881* (ILL). Liberty Co., Cleveland, *Ross 37* (ASTC). Limestone Co., Kosse, *Shinners 30975* (SMU). Lipscomb Co., Darrouzett, *Wallis 8447* (KANU, OKLA, SMU, TEX). Live Oak Co., George West, *Iwanicki & Wilkinson 7505–22* (TAES). Llano Co., Enchanted Rock, *Innes & Warnock 790* (GH). Lubbock Co., Lubbock, *Reed 3329* (TTC). Madison Co., Madisonville, *Gould 8558* (SMU, TAES, TEX). Marion Co., Jefferson, *Bebb 2658* (ILL, OKL, WIS). Maverick Co., Eagle Pass, *Cory 43853* (BH). McLennan Co., Waco, *Smith 404* (ILL, TEX, UARK). Milam Co., Cameron, *Wolff 3706* (F). Mitchell Co., Lavaca Nav. Co., *Pohl 4906* (ISC). Montague Co., Bowie, *Clement 71–76* (UARK). Montgomery Co., Conroe, *Raven & Gregory 19440* (DAO, DS, RSA, US). Nacogdoches Co., Nacogdoches, 1967, *Nixon s.n.* (ASTC). Navarro Co., Kerens, 1961, *Huggins s.n.* (ASTC). Newton Co., Newton *Cory 52645* (DS, FSU, MICH, MIL, NY, SMU, TAES, US). Nueces Co., Corpus Christi, *Tracy 9274* (CU, E, F, G, MIN, MO, MSC, NY, PENN, TAES, TEX, US, WIS). Orange Co., Mauriceville, *Nixon et al. 7107* (ASTC). Palo Pinto Co., Brazos, June, *Parks s.n.* (TEX). Panola Co., Gary, *Nixon & Sullivan 461* (ASTC). Parker Co., Weatherford, *Timmons 345* (SMU). Polk Co., Dallardsville, 1966, *Hendrix s.n.* (ASTC). Rains Co., Lone Oak, *Mosquin & Mosquin 5808* (DAO, DS). Refugio Co., Woodsboro, *Cory 54162* (GA, SMU, TEX). Robertson Co., Benchley, *Launchbaugh, Jr. 31* (TAES). Rusk Co., Henderson, *Reidel 44–12* (TEX). Sabine Co., Milam, *Chandler 272* (LTU). San Augustine Co., San Augustine, 1960, *Gonzalez s.n.* (ASTC). San Jacinto Co., Trinity River at jct. of Hwy 59, *Nixon 4421* (ASTC, NCU). San Patricio Co., Sinton, *Gould & Hycka 8015* (ARIZ, SMU, TEX, TTC, UC). San Saba Co., San Saba, *Wright 209* (TAES). Smith Co., Starville, *Lehio 15641* (ASTC). Somervell Co., Glen Rose, *Correll 15854* (GH, TEX). Starr Co., Rio Grande City, *Hanson 340* (MICH). Tarrant Co., Handley, *Ruth 444* (C, CM, NY, PH, TEX). Taylor Co., Abilene, *Tolstead 7507* (MICH, NY, TEX). Titus Co., Monticello, *Mahler 6488* (SMU). Travis Co., Austin, *Tharp 44140* (DS, DUKE, IND, NCU, NY, TAES, TEX, TTC, UC). Trinity Co., Apple Springs, 1960, *Davis s.n.* (ASTC). Tyler Co., Colmesneil, *Crider 31* (ASTC). Upshur Co., Big Sandy, *Mosquin & Mosquin 5464* (DAO, DS). Val Verde Co., 26 mi N of Langtry, *Warnock 47269* (SMU, TEX). Van Zandt Co., Grand Saline, *Tunnell 15* (SMU). Victoria Co., Victoria, 1900, *Eggert s.n.* (MIN). Walker Co., *Dixon 523* (CAS, POM, US, USF). Waller Co., Hempstead, *Trew, Jr., 199* (TEX). Washington Co., *Brackett 224* (GH, TEX). Wheeler Co., Shamrock, *Rowell 10082* (DS, RSA). Wilbarger Co., Electra, *Whitehouse 9816* (NY). Willacy Co., Yturria, *Runyon 2653* (POM, TEX). Williamson Co., Georgetown, *Wolcott 119* (TEX). Wilson Co., Stockdale, *Cory 54075* (GA, IND, KANU, SMU, TEX). Wise Co., Alvord, *Whitehouse 15009* (MICH, SMU). Wood Co., Hawkins, *Moody 110* (OKLA).—VERMONT: Chittenden Co., Burlington, 1896, *Saela s.n.* (VT). Rutland Co., Brandon, 1905, *Dutton s.n.* (VT).—VIRGINIA: Accomack Co., Chincoteague Island, *Gleason 8526* (NY). Amelia Co., *Lewis 1509* (UPI). Botetourt Co., Arcadia, *James N-169* (NCU). Brunswick Co., Seward Forest Area, 1940, *Lewis s.n.* (VPI). Chesterfield Co., 1890, *Colson, Jr. s.n.* (US). City of Virginia Beach: Shore Road, *Uttal 8696* (AUA, VPI). Culpeper Co., Rappahannock River on U.S. Rte 211, *Ahles & James 61470* (MASS, NCU). Dinwiddie Co., Petersburg, *Fernald & Long 10749* (GH, PH). Giles Co., foot of Peters Mtn, *Sharp & Fox s.n.* (GH, MO, PENN, PH, TENN). Goochland Co., Goochland, *James 6675* (NCU). Halifax Co., Danville, *Fosberg 15399* (PENN, US, WVA). Hanover Co., Atlee, *King 6399* (MICH, SMU, USF). Henrico Co., West Hampton, *Randolph & Merriman 282* (BH, CU, GH, UC). Henry Co., Piedmont, *Ramsey et al. 6794* (NCU, VPI). Isle of Wight Co., Lee's Mill, *Fernald et al. 14207* (PH). James City Co., Jamestown Island, *Artz 1100* (PENN, WVA). Louisa Co., along South Anna River, *Davis & Davis 10542* (WVA). Mathews Co., Blakes, *Foster 55* (ALU). Mecklenburg Co., Clarksville, *Fosberg 15440* (GH, PENN, US). Middlesex Co., Church View, *Leonard & Killip 540* (F, GA). New Kent Co., on Rte 627, S of jct. of Rte 627 & 1002, *Gillespie 52* (NCU). Nottoway Co., Blackstone, *James 6823* (NCU). Pittsylvania Co., jct. of Rte 706 &718, 1968, *Ruska & Waggoner s.n.* (NCU). Powhatan Co., Powhatan, *James 6247* (NCU). Prince Edward Co., Farmville, 1922, *Bright s.n.* (CM). Prince George Co., Camp Lee, 1919, *Bonar s.n.* (MICH). Roanoke Co., Salem, *Wood, Jr. 6231* (US). Rockingham Co., Grottoes, *Artz 891* (GH). Shenandoah Co., Devil's Hole Mtn, Alleghenies, *Allard 4631* (CM, F, GH, NY, US, VPI). Southampton Co., Franklin, *Heller 945* (BP, CU, DS, E, F, GH, MIN, MO, NY, PENN, PH, UC, US, VPI). Stafford Co., Falmouth, 1927, *Wiegand & Manning s.n.* (CU, GH, POM). Westmoreland Co.,

Lynch Point, *Iltis 920* (SMU).—West Virginia: Fayette Co., Prince Camp Site, 1968, *Phillips s.n.* (WVA). Grant Co., Petersburg, 1941, *Frye s.n.* (WVA). Hampshire Co., Yellow Springs, *Davis et al. 8164* (US). Kanawha Co., St. Albans, 1951, *Reed s.n.* (WVA). McDowell Co., Auawalt, 1962, *Music s.n.* (WVA). Mercer Co., East River, 1970, *Evans s.n.* (WVA). Pendleton Co., Smoke Hole, *Core 4339* (DS, WVA). Pleasants Co., Bull Creek, *Bartholomew P-124* (WVA). Pocahontas Co., Cass, 1929, *Core s.n.* (WVA). Raleigh Co., Terry, *Tosh 887* (WVA). Ritchie Co., North Hills, 1971, *Storestreet & Crane s.n.* (WVA). Summers Co., Barksdale, *Boone 452* (WVA). Wayne Co., Creek Cove, *Gilbert & Plymale 716* (DUKE, F, GA, GH, IA, ILL, MICH, MIN, MO, NY, OKL, PENN, PH, TENN, UMO, US, WIS, WVA). Wood Co., Washington, 1934, *Munchmeyer s.n.* (WVA).—Wisconsin: Columbia Co., Portage, *Rill 3083* (WIS). Green Lake Co., Princeton, 1968, *Soberalske s.n.* (WIS). La Crosse Co., Onalaska, 1955, *Peterson s.n.* (WIS). Monroe Co., Sparta, *Iltis & Neess 8958* (WIS). Sheboygan Co., Plymouth, 1903, *Goessl s.n.* (WIS). Waushara Co., Poygan, *Rill 3345* (WIS).

Specimens from Outside Natural Area. Argentina. Prov. La Pampa, Dept. Captial, Anguil, parque de la Estación Experimental del INTA, *Steibel 8532* (DUSS, SRFA); La Pampa, Dept. Chapaleufu, Banderalo, *Trioani & Prina 8226* (DUSS, SRFA); Chapaleufu, Ruta 188, between Meridiano V and Larroude, *Troiani & Steibel 7733.* Prov. San Luis, Dept. Pedernera, Ruta 148, 25 km S of Villa Mercedes, 450 m, *Anderson et al. 3121* (BAB).—Australia: New South Wales, Hunter River Distr., 1955, *Brown s.n.* (NSW).—Brazil: Rio Grande do Sul, near Porto Alegre, 1898, *Reineck s.n.* (LD).—Costa Rica: Zarcero, *Smith A-64* (F), *Weston 2106* (DS), *Weston 5686* (DS). Volcán Irazú, *Weston 2306C* (DS).—France: Seine et Oise, St. Saveur, 1919, *Despaty s.n.* (BAS); Basses-Pyrénées, Anglet, *Jallu 5331* (CAS, LISE); Nord, Dunkerque, *Bouly de Lesdain 63* (L); Bayonne, 1961, *Gavelle s.n.* (MA).—Germany: Freiburg, Baden-Württemberg, 1902, *Thellung s.n.* (BAS); Hamburg, 1912, *s.c., s.n.* (L), 1914, *Schmidt s.n.* (Z), *Schmitz 39* (BAS), 1900, *Schmidt 1806* (HBG), 1896, *Laban & Schmidt s.n.* (HBG); port of Neuss, Nordrheinwestfalen, 1915, *Bonte s.n.* (BAS); port of Düsseldorf, 1916, *Bonte s.n.* (BAS); Cologne (Niehl), 1931, *Hupke s.n.* (CAS); Mannheim, 1931, *Jung s.n.* (KR); Aggenmühle betw. Karlsruhe and Mühlburg, 1936, *Jung s.n.* (KR); Karlsruhe, 1936, *K. . . . [illegible] s.n.* (KR). Bavaria, Munich, 1929, *Dihm 3215* (M); Sachsen, Meissen, 1917, *Steifelhagen s.n.* (B); Berlin-Tegel, 1898, *Schulz s.n.* (B).—Italy: Piemonte, Torino, 1928, *Zola 14097* (FI); Firenze, 1952, *Chiarugi et al. s.n.* (FI); Liguria, Marina di Massa, 1932, *Baschant s.n.* (B); Torino, Castello del Villentino, 1928, *Effusa & Cresetti s.n.* (TO).—Japan. Spontaneous in Koishikawa Botanical Garden, Tokyo, *Makino 42843* (CAS, UC); Tokyo, 1909, *N.N. s.n.* (E); spontaneous in Botanical Garden, Tokyo, 1912, *Fox s.n.* (BM); Osaka, Sakai, *Makino 42844* (WVA); Osaka City, *Makino 42840* (CAN, CAS, DAO, UC, US, WVA); Hondo, 1959, *Furose s.n.* (A); Honshu, Isle Awaji, 1961, *Murata s.n.* (SMU, U); Honshu, near Kusuzaki, *Shimizu 13947* (U); Chiba Pref., Saka, Nishimisaki-mura, Awa-gun, *Kanai et al. 10092* (F, GH, K, UPS); Kyushu, Nichinan, Cape Toi, cult. Bot. Garden Munich Germany, *Merxmüller 26545* (M); Kyushu, Sendai City, *Boufford 19987* (MO).—The Netherlands. den Haag, 1935, *van Soest s.n.* (L); Deventer, 1916, *Kloos s.n.* (L); Erp, *Kern & Reichgelt 5202* (L); Leiden, *Tombe 515* (L); Loenen, 1924, *Jansen s.n.* (L); Nijmegen, *Kern & Reichgelt 5062* (L); Rhenen, 1963, *de Wilde s.n.* (L); Rotterdam, 1925, *Kloos s.n.* (L); Strijchen, 1941, *Jansen & Reichgelt 4506* (L); Waardingen, 1903, *Jansen & Wachter s.n.* (L); Weert, 1920, *Kloos s.n.* (L); Wormerveer, 1924, *Kloos s.n.* (L); Zaandijk, 1980, *Akkerman s.n.* (L); Zaanstad, 1976, *Akkerman s.n.* (L); Zeist, 1920, *Embden s.n.* (L).—Norway. Oslo, 1918, *Landmark s.n.* (O); Buvika, 1924, *Lyche s.n.* (O); Jölster, 1968, *Befring s.n.* (O).—Panama: Prov. Chiriquí, Cerro Punta, *Tyson 7093* (NA).—Paraguay. Dept. San Pedro, Alto Paraguay, Primavera, *Woolston 1024* (C, NY, S, SP, U, UC).—Portugal: Azores, Terceira, *Dansereau et al. 40* (NY).—Zimbabwe: Melsetter, Chipinga, *Eyles 8462* (K, SRGH).—South Africa. Cape: Stutterheim Div., Dohne Research Station, *Acocks 9016* (K); East London, Arnalinda Gardens, *Acocks 23488* (K); near Tokai, *Goldblatt 1435A* (MO). Lesotho: Maseru, 1700 m, *Williamson 21* (K), *Williamson 209* (K). Natal: Nottingham, *Galpin 9447* (K); Estcourt Exp. Station, *West 1129* (MO, PRE); Zululand, Ngome forester, *Gerstner 4501* (MO, PRE); Drakensberg Mts, Cathedral Peak, *Goodier 361* (K); Bergville, 1900 m, *Killick 2307* (K, M, PRE); Durban Distr., Insipingo, *Ward 3763* (PRE); Distr. Port Shepstone, Umzube, *Strey 6963* (K); Merrebank, *Bainath 259* (Univ. Coll. Durban); Wentworth Military Camp near Pietermaritzburg, *Strey 7783* (K, SRGH); Wembley near Pietermaritzburg, *Gordon-Gray 6358* (E, K, MO); Drakensberg Garden, *Jacobsz 2078* (MO). Oranje Free State: near Valsh River, Distr. Kroonstad, *Pont 669* (PRE, U); Bloemfontein, *Mostert 632* (PRE). Transvaal: Distr. Potschefstroom, *Loens 1345* (PRE); Saulspoort, Distr. Bethlehem, 1948, *Guratkin s.n.* (PRE); Experimental Garden, Pretoria, *Repton 5306* (K, PRE).—Spain: Prov. Barcelona, San Feliu de Codinas, 1938, *Garcías s.n.* (BC).—Sweden: Halland, Falkenberg, Valksvarnen, *Mattison 2959*

(CAS).—Switzerland: Zürich, 1913, *Thellung s.n.* (Z), 1913, *Beger s.n.* (B); Basel, 1922, *Becherer s.n.* (Z); Diessenhofen, 1922, *Spörri s.n.* (Z).—Taiwan: Ilan Co., Tungkan, Chuangwei, Hsiang, outlet of Lanyang River, 1984, *Hsing-fan Huang s.n.* (MO).—United Kingdom. Bermuda Islands: 1905, *Harshberger s.n.* (GH, MO, NY, PENN, US); Paget sand hills, *Brown & Britton 123* (NY, PH); Paynters Vale, *Brown 663* (C, GH, NY, PH, US); Smith's Parish, *Moore 2954* (CAS, GH, MICH); near Flatts, *Collins 105* (GH, NY), *Collins 385* (GH); St. Davis, *Brown 2090* (NY, PH); Harris Bay, 1921, *Bailey et al. s.n.* (BH); W of St. Georges I, *Taylor 49–1228* (MICH).—England: Calne, Keevil's Mill, V.C. 7, *Barton 700-2* (BM); North Somerset, Bedminster, 1922, *Sandwith s.n.* (K); Christchurch, Hampshire, 1923, *Lowne s.n.* (K); Pan Harbour, 1927, *Medlin s.n.* (K); Prestatyn Hintsh, V.C. 51, 1928, *Medlin s.n.* (K),; Lindwaite tip, SW-York, V.C. 63, *Webster 2771* (E, K); Maulden, Bedfordshire, V.C. 30, *Lousley 2180* (BM), *Dony 4371* (BM).—U.S.A. California: Fresno Co., Fresno Municipal Airport, 1959, *Buckaleu s.n.* (RSA). Los Angeles Co., Pasadena, *Munz 6127* (DS, POM, UC). Merced Co., Livingston, *Quibell 5804* (RSA). Orange Co., weed in Santa Ana Botanic Garden, *Balls 11190* (BM). Riverside Co., Banning, 1926, *Gilman s.n.* (POM). San Bernardino Co., Loma Linda, *Roos 3885* (RSA); Mentone, *Parish 6503* (LY). Santa Barbara Co., Goleta, 1958, *Pollard s.n.* (ARIZ, MIN, MO, RSA, TEX). Hawaii: Midway Atoll: Eastern Island, abandoned runway, *Herbst & Takeuchi 6404* (BISH); Sand Island, runway, *Herbst & Takeuchi 6396* (BISH). Maui: *Degener et al. 12593* (BH, CAS, CM, DS, GA, GH, ISC, KSC, MASS, MICH, MIL, MIN, MO, NY, PH, POM, TEX, UC); Kula, *Degener 28185* (BISH). Hawai'i: Kilauea, Old Volcano House, *Fagerlund & Mitchell 466* (BH, BISH); E of Waimea, *Wagner et al. 5544* (BISH).

Specimens intermediate between *Oenothera laciniata* and *O. grandis*. U.S.A. Oklahoma: Cherokee Co., 8.3 mi SE of Tahlequah, *Turley 74* (OKL). Cleveland Co., Norman, 1915, *Pitts s.n.* (OKL). Muskogee Co., 3 mi E of Fort Gibson on Hwy 10, *Wallis 3799* (OKL).—Texas: Kleberg Co., 6 mi E of Riviera on Hwy 771, *Gongora et al. 8815* (SMU 57% fertility). Webb Co., Fort McIntosh near Laredo, *Notzon 27* (DS, SMU 50% fertility). Willacy Co., Raymondville, *Wright 35* (MIN 55% fertility). Zapata Co., 2 mi E of Zapata, *Serna 23* (SMU 23% fertility); 4 mi E of Zapata on farm road 496, *Araiza 53* (DUKE, NA 68% fertility, SMU); 3 mi N of San Ygnacio, *Shinners 17655* (SMU 61% fertility); 10 mi S of San Ygnacio, *Gamez 70* (SMU, TAES); 10 mi N of San Ygnacio, Hwy 83, *Herrera & McCart 7696* (ARIZ, DUKE, MSC, TEX 50% fertility), *Herrera & McCart 7697* (VT).

Specimens intermediate between *Oenothera laciniata* and *O. humifusa*. U.S.A. Alabama: Baldwin Co., Fort Morgan Peninsula, *Iltis 21480* (DS); between Mobile & Spanish Fort, *Kral 55663* (VDB). Mobile Co., Mobile, *Deramus D854* (GH, UNA).—Florida: Alachua Co., 4 mi W of Gainesville, *Crosby 4842* (MO). Baker Co., Sapp, 1940, *West & Arnold s.n.* (FLAS). Brevard Co., Merritt Island, 1973, *Shuey s.n.* (FLAS). Citrus Co., Withlacoochee, *Genelle 1716* (ARIZ). Dixie Co., Shamrock, *Godfrey 65686* (DS, FSU). Duval Co., *Fredholm 5141* (GH, POM). Escambia Co., Pensacola, *Gander 7627* (SD). Franklin Co., Carrabelle, *Moldenke 26637* (BH, LL, WIS). Gadsen Co., near River Jct., 1938, *Abbott s.n.* (CU). Hardee Co., W of Zolfo Springs, *Ward A-34* (FLAS). Hernando Co., Weeki Wachee, *Genelle 647* (USF). Hillsborough Co., Dale Mabry, *Patman 1067* (GH, USF). Jackson Co., W of Chattachoochee at Apalachicola River, *Berkner 1162* (FLAS). Lake Co., vicinity of Eustis, *Nash 515* (E, GH, MICH, MSC, NY). Levy Co., Cedar Key, *Kral 2050* (FSU). Marion Co., Ocala near Summerfield, *Evans et al. 44470* (TENN). Nassau Co., Amelia Island, 1857, *Ward s.n.* (ILL). Okaloosa Co., 6.5 mi WSW of Crestview, *Shinners 26941* (SMU). Pasco Co., 1 mi N of Gower's Corner, *Cooley 5672* (GH, NY, USF). Pinellas Co., Clearwater, *Genelle 157* (USF). Polk Co., Lake Wales, *Demaree 49448* (DS). Putnam Co., 1.2 mi S of Welaka, *Wiggins 19918* (DS). St. John's Co., Hastings, 1940, *West & Arnold s.n.* (FLAS). Santa Rosa Co., Gulf Breeze, *Demaree 47279* (DS, SMU). Taylor Co., near Athena, *Beckner & D'Arcy 899* (FLAS). Volusia Co., New Smyrna Beach, *Murray 45528* (NCU, VDB, WIS). Walton Co., 3 mi S of Mossy Head, *Beckner 1365* (FLAS). Walluka Co., N of Panacea, St. Mark's Wildlife Refuge, *Trott 134* (DUKE, FSU, GH, SMS, VDB).—Louisiana: East Baton Rouge Pa., Baton Rouge, *Brown 3970* (LSU). Jefferson Davis Pa., Roanoke, *Thieret 22433* (DS). Latourde Pa., Godchaux Plant, *Ashbey 0020* (SMU). Lincoln Pa., 3 mi W of Ruston, *Chandler 206* (LSU). Livingston Pa., Albany, *Shinners 29610* (SMU). Orleans Pa., New Orleans, *Drummond 107* (NY). Plaquemines Pa., Saint Breton Island, *Stone 262* (MO). St. Charles Pa., W of New Sarpy Road, *Montz 496* (LSU). Terrebonne Pa., Mule lot, *Arceneaux 305* (CU). Vermilion Pa., 4.7 mi W of Kaplan, *Shinners 28059* (SMU).—Mississippi: Forrest Co., Leaf River near McCallum, *Rogers 6351-D* (VDB). Hancock Co., beach at Bay St. Louis, *Jones 11865* (MISS, NCU). Harrison Co., Gulfport, *Demaree 29005* (RSA, SMU). Jackson Co., Ocean Springs, *Pollard 1020* (F, POM, US).—North Carolina: Brunswick Co., 3 mi S of Hwy 174 on Hwy 40, *Bell 11480* (NCU, WVA). Columbus Co., Old Dock near

Waccamaw River Bridge, *Watt 733* (DUKE). New Hanover Co., 6 mi N of Wilmington on Hwy 421, *Wilbur 17838* (DUKE). Pender Co., Surf City, *Ahles 23459* (DAO, ISC, NCU).—South Carolina: Beaufort Co., Hunting Island State Park, *Bell 2477* (NCU). Georgetown Co., Huntington Beach State Park, *Iltis 33159* (UC); 8 mi S of Pawley's Island, *Iltis 23182* (UC, WIS).

Specimens intermediate between *Oenothera laciniata* and *O. mexicana.* U.S.A. Texas: Brooks Co., 12 mi S of Falfurrias, *Lundell 10805* (MICH, POM, SMU). Kenedy Co., El Toro Island, mud flats of the Laguna Madre, *Tharp 49138* (GH, OKL). Nueces Co., Port Aransas, Mustang Island, *Mahler 5348* (NA, SMU 34% fertility, TEX). San Patricio Co., 4 mi SE of Edroy on Hwy 9, *Whitehouse 18137* (MICH 20% fertility, NY 10% fertility, SMU 33% fertility, TEX).

Oenothera laciniata is an autogamous, permanent structural heterozygotic species. It is easily the most widespread species of subsect. *Raimannia,* and one of the most widely naturalized species in the genus. Its ecological amplitude appears to be much greater than that of O. *humifusa,* the other permanent structural heterozygote of ser. *Raimannia,* which is restricted to the dunes of the Atlantic coast. Considering the large geographical range of *O. laciniata,* it is not surprising that the morphological variation is correspondingly great.

The most common morph of *O. laciniata* has deeply divided leaves, petals 5–10 mm long, and the calyx lobes subterminal. All of the cultivated strains were the common morph except those noted below in the discussion of other morphs. There are also other major morphs that are especially striking; they are, however, connected by transitional forms with the common morph.

A form of *O. laciniata* with less deeply divided leaves with broad terminal lobes and with minute free sepal tips that are coherent in bud so that the bud is acuminate occurs in southeastern Texas, within the range of *O. falfurriae* and *O. mexicana.* The size of flowers corresponds with the common form. Plants of this type were grown from the seeds of *Shaw & Allred 2016, 2018,* and *2020,* as well as *Jones s.n.* in 1974, from Padre Island (DUSS 33, 339 & 343 from *2016;* 80–318 & 324 pro parte from *2018;* 80–310 from *2020*). Plants of the more typical form were also among the seeds from Shaw and Allred (all numbers not mentioned above) and transitional forms (79–121 & 128 from *2016;* 79–139, 80–327, 328, 329 & 331 from *2021*). Since it is not possible to make a discreet morphological separation between the common form and the form with minute sepal tips, and since these forms occur sympatrically, they are not formally recognized here. *Oenothera mexicana* apparently was involved in the origin of this form. This hypothesis is based on the sympatric occurrence of *O. mexicana* with this form, but, more importantly, because the only characters that differentiate this form from the most common one, namely short sepal tips, nearly erect floral tubes, and dense spreading pubescence, are also shared with *O. mexicana.* This hypothesis has not been evaluated experimentally, however, because of the difficulties in crossing *O. mexicana* and *O. laciniata.*

A third form, occurring in the Atlantic coastal region from Virginia to Louisiana, is distinguished by noticeably larger petals (1.4–2.3 cm long), leaves less deeply incised, and thick, often spreading sepal tips. Plants of this form are usually more prostrate and are apparently short-lived perennials. We have cultivated material of this form from Alabama, Florida, Mississippi (*Rogers 6668-A*), North Carolina (Kew 0209–12), and Virginia (1974, *Straley s.n.*). The formation of this prostrate form presumably has involved hybridization with *O. drummondii* or *O. humifusa,* with which it shares the prostrate habit, appressed pubescence, and the tendency towards a perennial growth habit. This hypothesis is supported by experimental hybridization; for example, in crosses between *O. grandis* (Kansas, Brenham, *Brooks 6616*) and *O. drummondii* (Texas, Mustang Island) some progeny (DUSS cult. no. 76–213) strongly resemble the prostrate form of *O. laciniata* in habit and leaf lobing. Likewise, the crosses between *O. humifusa* (Ocean City) and the *O. grandis* strain mentioned above resulted in a similar phenotype except for the different flower size (DUSS cult. no. 76–422).

Both complexes of the common form of *O. laciniata* probably were derived from *O. grandis*. This is suggested by our own observations as well as by allozyme studies and crossing results of Ellstrand and Levin (1980b, 1980c). They state that *O. laciniata* probably arose several times as an "interpopulational hybrid" from *O. grandis* independent of time and place. *Oenothera laciniata* differs from other species of subsect. *Raimannia* in that apparently only one chromosome complex is transmitted by the egg cell (α) and the other by the pollen (β) (Ellstrand & Levin 1980a, 1980b).

In summary, the analysis of the forms of *O. laciniata* reveals that the following genome combinations or at least parts of the genomes apparently were involved: *O. grandis–O. grandis, O. grandis–O. mexicana,* and *O. grandis–O. drummondii.*

10. Oenothera drummondii Hooker, Bot. Mag. 61: *t. 3361*. 1834. *Raimannia drummondii* (Hooker) Rose ex Sprague & Riley, Bull, Misc. Inform. 1921: 200. 1921. *Oenothera sinuata* race *humifusa* var. *drummondii* (Hooker) H. Léveillé, Monogr. Onoth. 351. 1909.—Type: Grown at the Glasgow Botanical Garden, seeds collected by *Drummond* at Rio Brazos, Texas [probably near Freeport, Brazoria County] (holotype: E!; isotypes: G! GH! K!).

Erect to procumbent annual to perennial herbs from a taproots up to 2 cm in diameter, usually not forming rosettes but growing soon after germination from only a few basal leaves, non-flowering lateral branches often with a terminal rosette of crowded small leaves; stems 1–5 dm long, stiff, green, sometimes flushed with red, simple or much-branched, densely strigillose or sometimes also villous, becoming glandular-puberulent in upper part. Basal leaves 5–14 cm long, 1–2 cm wide, narrowly oblanceolate to elliptic, remotely shallowly dentate to almost entire, apex acute, gradually or abruptly narrowed to the petiole; cauline leaves 1–8 cm long, 0.5–2.5 cm wide, narrowly elliptic to broadly elliptic or narrowly obovate to broadly obovate, remotely shallowly dentate to subentire, rarely lyrate, apex acute to rounded, gradually or abruptly narrowed to the short petiole; bracts 0.8–5.5 cm long, 0.4–1.8 cm wide, narrowly elliptic to broadly elliptic or narrowly oblanceolate to oblanceolate, remotely and bluntly shallowly dentate to subentire, apex acute to obtuse, base narrowly cuneate to rounded, short-petiolate or sessile; leaves and bracts grayish green, densely strigillose, rarely also glandular-puberulent. Inflorescence lax, usually with lateral branches. Usually one flower per spike opening each day near sunset. Floral tube 2–5 cm long, 1.5–2 mm in diameter, densely strigillose to villous, sometimes also scattered to sparsely glandular-puberulent. Mature buds 5–11 mm in diameter at the base, narrowly lanceoloid or narrowly ovoid to ovoid. Sepals 1.3–3.3 cm long, green to yellowish, often flushed with red and striped red at the margins, sometimes also red-maculate, pubescence like that of floral tube, free sepal tips 0.3–3 mm long, erect and appressed in bud, strigillose. Petals 2–4.5 cm long, 2.5–5.5 cm wide, yellow, very broadly obovate, truncate to emarginate. Filaments 1–2.3 cm long; anthers 4–12 mm long, pollen ca. 90–100% fertile. Ovary

1.3–2.5 cm long, ca. 1.5 mm in diameter, densely strigillose to villous, sometimes also sparsely glandular-puberulent. Style 3.5–7.5 cm long, the exserted part 1.5–3.5 cm long; stigma elevated above the anthers at anthesis, lobes 3–10 mm long. Capsule 2–5.5 cm long, 2–5 mm in diameter, cylindrical, pubescence same as ovary. Seeds 1.1–2 mm long, 0.5–0.9 mm in diameter, ellipsoid to broadly ellipsoid, rarely suborbicular, brown, sometimes with darker flecks, the surface pitted. Self-compatible, modally outcrossing. Chromosome number: $n = 7$ (7_{II}*, ⨀4 and 5_{II}**, ⨀6 and 4_{II}***, or ⨀8 and 3_{II}*** at meiotic metaphase I).

Oenothera drummondii has large flowers and more or less dense, appressed pubescence. It is self-compatible and an annual to a short-lived perennial that persists for several years. Both the perennial habit and the often prostrate stems probably represent adaptations to its coastal habitat. *Oenothera drummondii* occurs only in this narrow ecological range even in places where it is naturalized. *Oenothera drummondii* is here subdivided into two subspecies, one of which occurs along a long stretch of the Atlantic coast and the other in a single small region of the Pacific coast of Baja California.

Although *O. drummondii* is self-compatible, its large flowers and elevated stigma suggest that it is largely outcrossing; however, it is capable of self-pollination in the absence of pollinators. Gregory (1963, 1964) observed the hawkmoth, *Hyles lineata,* pollinating *O. drummondii* at Aransas Pass, San Patricio Co., Texas. *Lasioglossum texanum* was also observed there removing pollen from *O. drummondii,* but it does not appear to be an effective pollinator.

The separation of *O. drummondii* subsp. *thalassaphila* from subsp. *drummondii* depends on a combination of characters, since there is no single morphological feature that separates them clearly. In addition to those characters mentioned in the key, several other features are modally differentiated. *Oenothera drummondii* subsp. *thalassaphila* always grows for several years, as is demonstrated by the consistent presence of nonflowering shoots and large taproots in the older plants. In contrast, *O. drummondii* subsp. *drummondii* is basically an annual, seldom overwintering for a second season; it usually has only a few nonflowering shoots or none, and the development of its taproot is considerably weaker than in subsp. *thalassaphila.* In general, the habit of subsp. *drummondii* is more upright than that of subsp. *thalassaphila,* which has prostrate to ascending stems. In addition, the calyx of subsp. *thalassaphila* often has red spots and lacks glandular hairs, whereas the calyx of subsp. *drummondii* only rarely has reddish spots and is often glandular-puberulent.

Key to the Subspecies of Oenothera drummondii

1. Floral tube 2.5–5 cm long; sepal tips 1–3 mm long; capsule 2.5–5.5 cm long, 2–3 mm in diameter; seeds 1.1–1.7 mm long, 0.5–0.8 mm in diameter; cauline leaves up to 8 cm long. 10a. *O. drummondii* subsp. *drummondii.*
1. Floral tube 2–3.5 cm long; sepal tips 0.3–1 mm long; capsule 2–4 cm long, 2.5–5 mm in diameter; seeds 1.5–2 mm long, 0.7–0.9 mm in diameter; cauline leaves up to 4.5 cm long. 10b. *O. drummondii* subsp. *thalassaphila.*

10a. Oenothera drummondii subsp. **drummondii.**

Oenothera littoralis Schlechtendal, Linnaea 5: 556. 1830 [description]; Linnaea 12: 268. 1838 [name]. *Raimannia littoralis* (Schlechtendal) Rose ex Sprague

& Riley, Bull. Misc. Inform. 1921: 201. 1921.—TYPE: MEXICO. Veracruz: sandy seashore between Tecoluta and Nautla, Feb 1829, *Schiede 532* (not *Ehrenberg* as stated by Schlechtendal) (holotype: HAL-43355!).

Oenothera sinuata race *humifusa* var. *drummondii* subvar. *helleriana* H. Léveillé, Monogr. Onoth. 351, 357. 1909. *Oenothera drummondii* var. *helleriana* H. Léveillé, Monogr. Onoth. *t.* opposite p. *356.* 1909 [this apparently an error for subvar. *helleriana* in the caption].—TYPE: U.S.A. Texas: Nueces Co., Corpus Christi, 23–30 Mar 1894, *Heller 1512* (lectotype, designated by Lauener, 1972: E!; isolectotypes: ARIZ! BM! CU! F! G! GH! ISC! K! MASS! MICH! MIN! MO! MSC! NY! POM! SMU! UC!).

Erect to procumbent annual to short-lived perennial herbs from taproots not more than 1 cm in diameter; stems 2–5 dm long, usually without non-flowering shoots; stems, leaves, ovary and capsule exclusively strigillose or strigillose to villous and the stem glandular-puberulent in upper part. Cauline leaves 1–8 cm long, 0.5–2.5 cm wide; bracts 1–5.5 cm long, 0.4–1.8 cm wide. Floral tube 2.5–5 cm long. Buds 7–11 mm in diameter at base. Sepals 2–3 cm long, sepal tips 1–3 mm long. Petals 2.5–4.5 cm long, 3–5.5 cm wide. Anthers 5–12 mm long. Style 5–7.5 cm long, the exserted part 2.5–3.5 cm long; stigma lobes 5–10 mm long. Capsule 2.5–5.5 cm long, 2–3 mm in diameter. Seeds 1.1–1.7 mm long, 0.5–0.8 mm in diameter. Self-compatible, modally outcrossing. Chromosome number: $n = 7$ (7_{II}*, ⊙4 and 5_{II}**, ⊙6 and 4_{II}***, or ⊙8 and 3_{II}**** at metaphase I). Fig. 10.

Phenology. Flowering throughout the year.

Distribution (Fig. 11). Along the Atlantic coast on dunes and open sandy places from New Hanover Co., North Carolina, south to Campeche, Mexico. Collections at inland localities in Bexar and Dallas counties, Texas, and Henderson Co., North Carolina, presumably represent introductions. This subspecies has been reported in the literature from outside of North America as follows: Australia (Beadle et al. 1972; Blackall & Grieve 1974); Israel (Zohary 1972); and Peru (Dietrich 1977).

SPECIMENS FROM CULTIVATED PLANTS. U.S.A. SOUTH CAROLINA: Charleston Co., Sullivan's Isle, *Cleland 468* [cult. no. 76–727**] (DUSS, M, MO); Palm Island near Charleston, 1974, *Behnke s.n.* [cult. no. 75–1450*] (DUSS, M, MO); Charleston, *Raven 20457* [cult. no. 77–531*] (DUSS, M, MO).—TEXAS: Nueces Co., Padre Island, 1974, *Whistler s.n.* [cult. no. 75–1446**] (DUSS, M, MO); Mustang Island, 1974, *Jones s.n.* [cult. no. 76–723****] (DUSS, M, MO).—ISRAEL: Tel Aviv, 1974, *Eisikowitsch s.n.* [cult. no. 75–1442*] (DUSS, M, MO).—MEXICO. VERACRUZ: Palma Sola, 1974, *Rzedowski s.n.* [cult. no. 75–1440***] (DUSS, M, MO).—UNITED KINGDOM. BERMUDA ISLAND, 1974, *Irwin s.n.* [cult. no. 75–1451*, 75–1459**] (DUSS, M, MO).—CZECHOSLOVAKIA: Praha, 1840 (PRC).—FRANCE. Paris, 1835 & 1841 (FI-Herb. Webb), 1848 (DS) (as *O. douglasii*).—GERMANY. Munich, 1839, *Zuccarini s.n.* (M), 1846, *Kummer s.n.* (M); Celle, 1866, *herb. Nöldeike* (BREM).—ITALY. Firenze, 1857 (FI).—PORTUGAL. Azores, 1868 (BM).—U.S.A. PENNSYLVANIA: Philadelphia (Hort. cantabr.), 1845, *Gray s.n.* (F, GH, K).

REPRESENTATIVE SPECIMENS. MEXICO. TAMAULIPAS: Playa Washington, 38 km E of Matamoros, 9 km S sobre la playa, *Medrano 521* (MEXU); vicinity of La Barra, 8 km E of Tampico, *Palmer, E. 282* (GH, NY).—VERACRUZ: Boca Andrea, Alto Lucero (19°47′N, 96°25′W), *Dorantes 397* (MEXU, MO); cult. from *Dorantes & Acosta 2679* from Laguna de la Mancha (19°34′N, 96°23′W) (MO); Playa Casitas (20°19′N, 96°49′W), *Gómez-Pompa & Nevling 1256* (MEXU, MO, NY); 13 km NW of Alvarado along Hwy 180 (18°50′N, 95°51′W), *Hansen & Nee 7640* (MO); Medanos frente a Laguna de Tamiahua rumbo a cabo Rojo, *Lot 1299* (GH, MEXU); Tecolutla, *Rzedowski 20008* (MO); Faro del Punta Delgada entre Palma Sola & Punta del Morro, *Soto & Horvitz 9* (MEXU); Nautla, *Ventura 12553* (MEXU, MO); ca. 6 mi E of Tuxpán, *Ward 7836* (NY).——U.S.A. FLORIDA: Brevard Co., S end of Merritt's Island, *Small &*

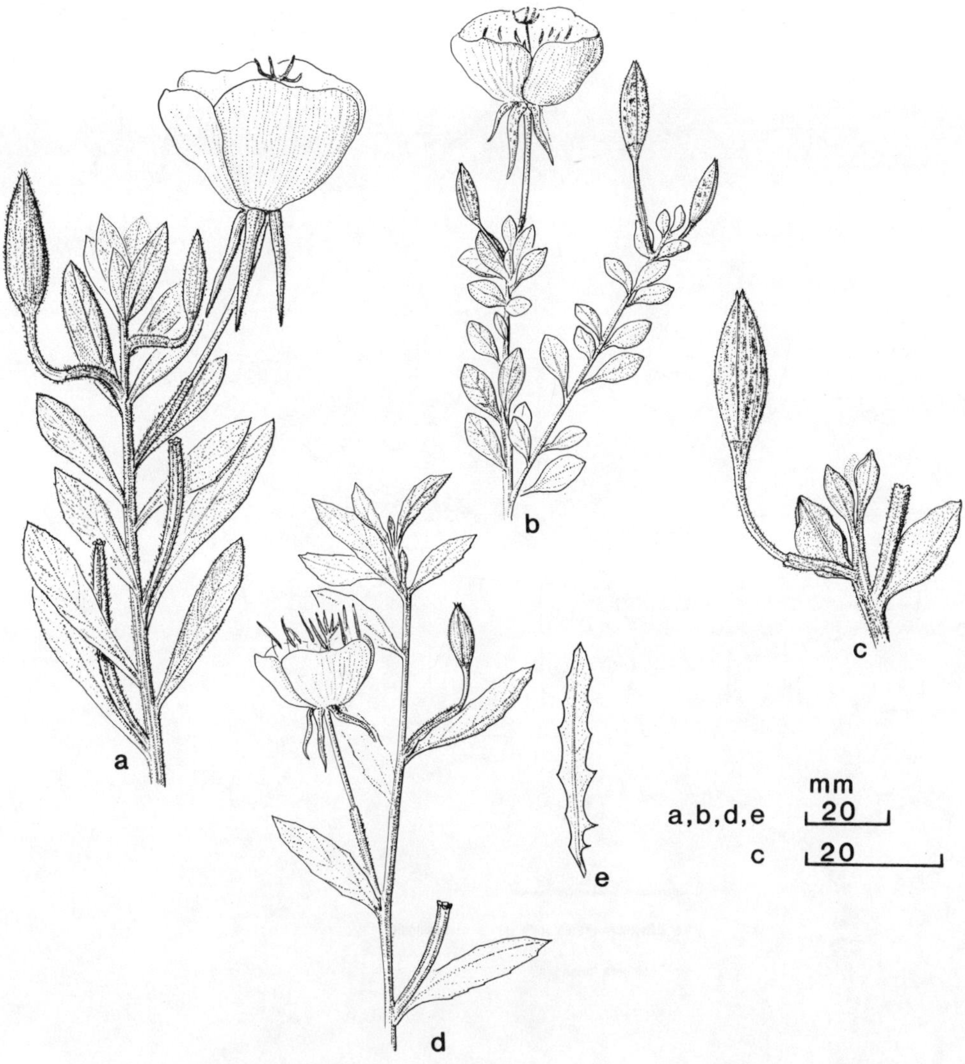

FIG. 10. *Oenothera drummondii* and *O. humifusa. O. drummondii* subsp. *drummondii* (*Traverse 1167;* MO; *Taylor & Taylor 13255,* Texas, Jefferson Co.; MO): a. Inflorescence. *O. drummondii* subsp. *thalassiphila* (*Constance 3183;* MO): b. Inflorescence. c. Bud with bracts. *O. humifusa* (*Ellstrand s.n.* in 1978, cult. no. 79–92; MO; *Binding s.n.* in 1977, cult. no. 79–93; MO): d. Inflorescence (*Ellstrand*). e. Bract (*Binding*).

DeWinkeler 9465 (NY). Dade Co., near Naranha, *Small 8526* (NY). Volusia Co., Ormond, near canal, 1903, *Purdie s.n.* (GH).—LOUISIANA: Cameron Pa., Holly Beach, *Correll & Correll 9603* (DUKE, F, GH, LSU, NA, ND, NY). Jefferson Pa., Grand Isle, *Brown 2012* (LSU).—NORTH CAROLINA: Henderson Co., Hendersonville, *Campbell s.n.* (NCU). New Hanover Co., Fort Fisher Beach, 1953, *Henry & Beer s.n.* (CM).—SOUTH CAROLINA: Charleston Co., Sullivan's Island, *Leonard & Radford 2169* (ALU, ARIZ, ASC, AUA, B, C, CM, DS, E, FLAS, FSU, GA, GH, ISC, KANU, KE, MASS, MIN, MIS, MO, NCU, NO, NY, OKLA, PAM, RSA, SIU, SMS, SMU, TENN, TEX, UC, USF, VDB, VPI, WIS, WVA).—TEXAS: Aransas Co., ferry to Port Aransas, *Gregory 416* (CAS, DAO, DS, NCU, RSA, UC). Bexar Co., San Antonio, *Ruth 1556* (POM). Brazario Co., Freeport, *Cory 51044* (GH, SMU). Calhoun Co., Port O'Connor, *Hatch 2095* (TAES). Cameron Co., Boca Chica, *Traverse 1167* (F, GH, MO,

FIG. 11. Distribution of *Oenothera drummondii* subsp. *drummondii* and *O. humifusa*.

SMU, TEX, US). Chambers Co., S of High Island, *Mahler 5163* (MSC, OKL, SMU). Dallas Co., Dallas, 1926, *Draper s.n.* (NY). Fort Bend Co., *Palmer 5007* (P). Galveston Co., Galveston, *Tracy 9222* (CU, E, F, G, GH, MIN, MO, MSC, NY, PENN, TAES, TEX, US, WIS), *Lindheimer 69* (FR), *Lindheimer 53* (FI, K, P). Harris Co., Morgans Point, *Palmer 11965* (GH, MO, NY, US). Jefferson Co., *Taylor & Taylor 13255* (MO); 17.7 mi E of Chambers Co. line in Hwy 87, *Traverse 968* (ASTC, F, MO, TEX, US). Nueces Co., Corpus Christi Bay, *Palmer 343* (AC, K, MO, NA, NY, P, POM, VT). Refugio Co., *Palmer 9248* (P). San Patricio Co., between Port Aransas & Aransas Pass, *Munz & Gregory 23450* (DAO, RSA, UC). Victoria Co., McFaddin Beach, *Crockett 49A* (TEX). Willacy Co., Port Mansfield, *Traverse 1188* (SMU, TEX). County unknown: Rio Brazos, *Drummond 26* (K), 1833, *Drummond s.n.* (K).

Representative Specimens from Outside Natural Area. Argentina. Buenos Aires: La Paternal, 1919, *Morfino 1475* (AMD).—Australia. New South Wales: Caloundra, 1943, *Clemens s.n.* (RSA), *Thorne 20166a* (L, RSA); Coolum Beach, 1945, *Clemens s.n.* (GH, MICH); Moreton Island, 1924, *White s.n.* (US); Redcliff, Moreton Bay, 1926, *White s.n.* (US); Coffs Harbor, *Meebold 3444* (M), 1941, *Comish s.n.* (NSW); Nowra, 1937, *Rodway s.n.* (NSW); Port Stephens, Nelson Bay, 1961, *Evans s.n.* (NSW), 1965, *Litgow s.n.* (NSW); Stockton, *McDonald 5104* (NSW); Lord Howe Island, Kindy Point, 1965, *Mears s.n.* (NSW); Newcastle, *Briggs 3254* (K); Mosman Park, *Alplin 1072* (PERTH).—Queensland: Redcliff, Moreton Bay, 1926, *White s.n.* (BM, K, MEL, US); Bribie Island, Moreton Bay, *Hubbard 2694* (K, L); Amity Point, Stradbrooke Island, *Hubbard 2329* (K), *White 6745* (NSW), *Mekee 9495* (NSW); E. Fraser Island, Wide Bay, *Baxter 940* (K).—Western Australia: W coast between Moore River and Cape Leeuwin, *Sauer 3411* (WIS); Fremantle, 1947, *Meadly s.n.* (PERTH), 1973, *Grigy s.n.* (PERTH); Cattesbee, 1952, *Johnston s.n.* (PERTH); Swanbourne, 1952, *O'Grady s.n.* (PERTH).—China: Fukien, Amoy Island, *Chung 1524* (E, K, UC, W); Double Island, near Swatow, 1900, *Dalziel s.n.* (E,K), *Gressitt 1772* (GH); Amoy, *Hom 19168* (A), *Chung 1524* (E, K, UC); Amoy Island, Steward 3084 (A, UC); Kwangtung, *Tsoong 4337* (A); Canton, Macao, *Hance 4375* (BM, K), *Sampson 244* (BM).—Egypt: Giza, Orman Gardens, *Simpson 5056* (K); betw. El-Arish and Rafah, 1965, *Boulos s.n.* (K).—Iraq: Abu Ghraib near Erbil, *Barkley 5570* (K).—Israel: 5 mi S of Tel-Aviv on coast, *Curle 102* (E); Bet Tam near Jaffa, *Davis 3730* (E, K); Acre, on dunes from Hanfa to Acre, *Davis 3832* (E, K); Acre, *Davis 4100* (E, K); Jaffa, *Dinsmore B1074* (A, B, G, L, MIN); Tel Aviv, *Eig et al. 269* (AMD, BC, C, CAS, DAO, E, FI, G, GH, HUJ, K, KRAM, MA, P, U, US, W); Akko (Acre), *Kramer 4602* (U); Jaffa, 1902, *Meyer s.n.* (E); Hula Plain, 1967, *Zohary s.n.* (HUJ); Sharon Plain, Netauya, 1971, *Zohary s.n.* (HUJ); Philistean Plain, Bat-Yam, 1947, *Zohary s.n.* (HUJ); Acre Plain, Qiryat-Yam, *Leinkram 20252* (HUJ).—Morocco: Rio Martin near Tetouan, *Font Quer 460* (B, BM, MA).—South Africa: Transvaal, Barberton, 1960, *Duncan s.n.* (K); Cape, Port Elisabeth, *Theron 1146* (K); Natal, Inanda, *Klatmough 492* (K).—Spain: Rota, *Millward 74* (BM); Guipuzcoa, Zarauz, 1969, *Rivas Goday s.n.* (MA).—United Kingdom. Bermuda Island: Warwick Camp, *Rendle 682* (BM); Bernauda, 1873, *Meseley s.n.* (BM).

Specimens intermediate between *Oenothera drummondii* subsp. *drummondii* and *O. grandis.* Mexico. Tamaulipas: Tampico, *Orcutt 5536* (MO); San José, *LeSueur 320* (F, TEX); vicinity of Tampico, 1910, *Palmer, E. s.n.* (US); vicinity of La Barra, 8 km E of Tamaulipas, *Palmer, E. 255* (CAS, G, GH, MO, NY, US); Tampico, *Rutter et al. 471* (U).—Veracruz: ca. 6 mi E of Tuxpán, *Ward 7837* (FLAS, MICH).

Specimens intermediate between *Oenothera drummondii* subsp. *drummondii* and *O. laciniata.* U.S.A. Louisiana: Cameron Pa., 2 mi E of Holly Beach, *Thieret 8839* (RSA).—Texas: Fort Bend Co., Arcola, *Palmer 5097,* pro parte (CU). Galveston Co., Galveston Ferry, *Olney 8* (DS, OKL). Jefferson Co., 21 mi NE of High Island, *Cory 50996* (DS, GH, MICH, NY, SMU, TEX, US). Jim Hogg Co., 5.5 mi S of Agua Nueva, *Johnston 54141* (TEX). Kenedy Co., El Toro Island, *Tharp 49138* (F). Matagorda Co., S of Matagorda, 6.5 mi S of bridge, *Whitehouse 20876* (SMU).

Besides *O. laciniata, O. drummondii* subsp. *drummondii* is the only member of ser. *Raimannia* that has successfully naturalized in suitable habitats in Africa, Asia, Australia, and Europe.

10b. Oenothera drummondii subsp. **thalassaphila** (Brandegee) W. Dietrich & W. L. Wagner, Ann. Missouri Bot. Gard. 74: 150. 1987. *Oenothera thalassaphila* Brandegee, Univ. Calif. Publ. Bot. 10: 185. 1922. *Oenothera drummondii* var. *thalassaphila* (Brandegee) Munz, Amer. J. Bot. 22: 651. 1935.—TYPE: MEXICO. Baja California Sur: San José del Cabo, 12 Mar 1892, *Brandegee 218* (lectotype: designated by Munz, 1935: UC-107674!).

Ascending to procumbent perennial herbs, old individuals with taproots up to 2 cm in diameter; stems up to 3 dm long, often with non-flowering shoots bearing terminal rosettes of crowded small leaves; stems, leaves, ovary and capsule exclusively strigillose. Cauline leaves 1–4.5 cm long, 0.7–1.5 cm wide; bracts 0.8–3.5 cm long, 0.4–1.4 cm wide. Floral tube 2–3.5 cm long. Mature buds 5–8 mm in diameter. Sepals 1.3–2.5 cm long, sepal tips 0.3–1 m long. Petals 2–3.5 cm long, the exserted part 1.5–2.3 cm long; stigma lobes 2.5–4 mm long. Capsule 2–4 cm long, 2.5–5 mm in diameter. Seeds 1.5–2 mm long, 0.7–0.9 mm in diameter. Self-compatible, modally outcrossing. Chromosome number: n = 7 (7_{II}* at meiotic metaphase I). Fig. 10.

Phenology. Flowering probably throughout the year.

Distribution (Fig. 12). Restricted to dunes of coastal southern Baja California Sur, Mexico.

SPECIMENS FROM CULTIVATED PLANTS. MEXICO. BAJA CALIFORNIA SUR: 11 km E of Cabo San Lucas, *Carter 5848* [cult. no. 75–1444*] (DUSS, M, MO); sand dunes near Cabo San Lucas, *Breedlove 13444* [cult. no. 78–217*] (DUSS, M, MO).

ADDITIONAL SPECIMENS EXAMINED. MEXICO. BAJA CALIFORNIA SUR: 3.2 km W of San José del Cabo (ca. 23°05′N, 109°35′W), *Carter 5848* (MO); Puerto de Bahia de los Muertos, SE of La Paz (23°55′N, 109°45′W), *Chambers 850* (DS, MEXU, SD, UC); Cabo San Lucas (22°53′N, 109°54′W), 1947, *Constance 3183* (DS, F, GH, K, LL, MEXU, MICH, MO, NY, UC, US), *Barcley 3180* (BM); 29 mi N of Cabo San Lucas, *Hastings & Turner 64-354* (ARIZ, DS, SD); Todos Santos (23°27′N, 110°13′W), *Jones 24098* (CS, DAO, DUKE, F, GH, MICH, MO, NA, NY, PH, POM, RSA, SD, SMU, TEX, UC, US); 8 km S of Pescadero (near 23°21′N, 110°11′W), *Moran 7016* (CAS, DS, SD); SW edge of Los Frailes (23°22′N, 109°26′W), *Porter 307* (CAS, DS, RSA); Rancho de la Ventana, western end of Bahia de la Ventana (24°05′N, 109°55′W), *Wiggins 14423* (ARIZ, CAS, DS, TEX, UC); near resort of Buena Vista, SE of La Paz and El Triunfo (near 23°47′N, 110°08′W), *Wiggins 14692* (DS, GH, K).

The plastome of this geographically disjunct subspecies is differentiated from that of subsp. *drummondii:* hybrids between subsp. *thalassaphila* and subsp. *drummondii* or *O.grandis* are pale green or variegated. Hybrids between *O. drummondii* subsp. *drummondii* and *O. grandis* are a normal green, which suggests that they have similar plastid types.

Because of its small range and its occurrence at sites heavily visited by tourists, *Oenothera drummondii* subsp. *thalassaphila* may be considered endangered.

11. Oenothera humifusa Nuttall, Gen. N. Amer. Pl. 1: 245. 1818. *Oenothera sinuata* var. *humifusa* (Nuttall) Torrey & A. Gray, Fl. N. Amer. 1: 494. 1840. *Raimannia humifusa* (Nuttall) Rose ex Britton & A. Brown, Ill. fl. n. U.S., ed. 2, 2: 597. 1913. *Oenothera sinuata* race *humifusa* (Nuttall) H. Léveillé, Monogr. Onoth. 351. 1909.—TYPE: U.S.A. Florida: [presumably Nassau Co.], sea beach near Cumberland Island, [Georgia], on the dry sandy beach, *Baldwin 883* (holotype: PH-916111!).

Oenothera niveifolia Gandoger, Bull. Soc. Bot. France 65: 27. 1918.—TYPE: U.S.A. Florida: Escambia Co., Perdido, 25 May 1903, *Tracy 8719* (holotype: P!; isotypes: BM! CU! E! F! G! GH! MIN! MO! MSC! NCU! NY! PENN! TAES! UMO! US! WIS!).

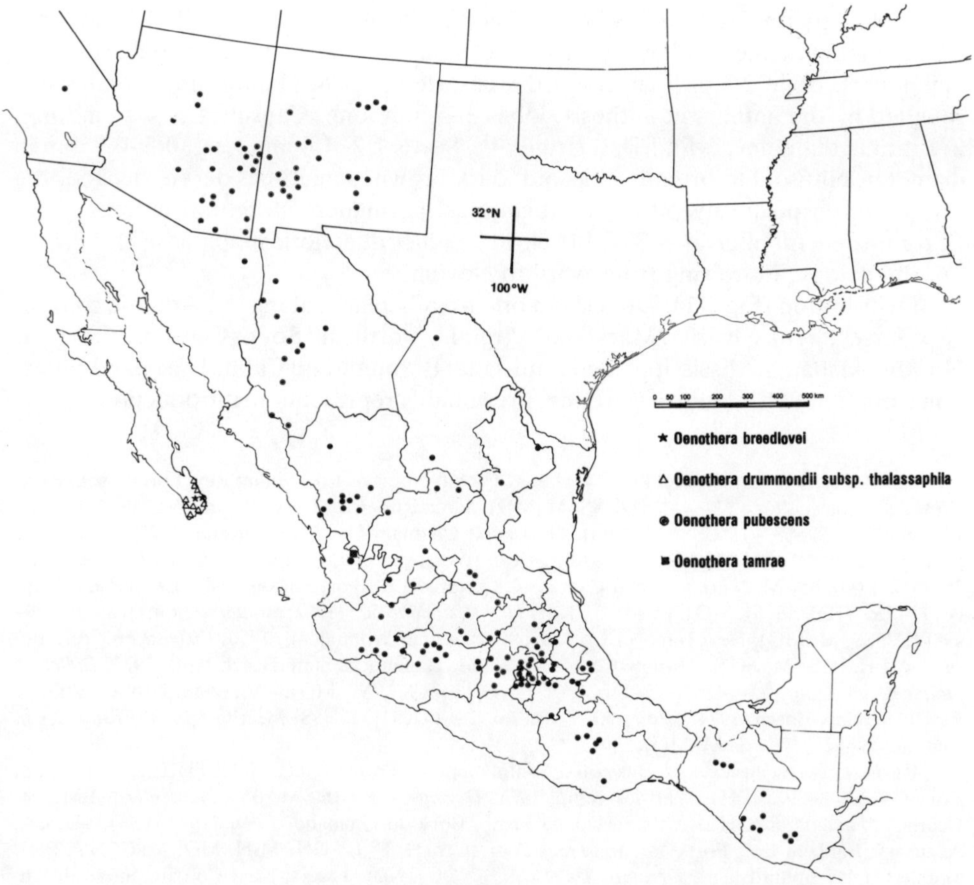

FIG. 12. Distribution of *Oenothera breedlovei, O. drummondii* subsp. *thalassaphila, O. pubescens,* and *O. tramrae.*

Erect to procumbent annual or short-lived perennial herbs from definite rosettes, sometimes with only a few basal leaves; stems 1–5 (–9) dm long, stiff, simple to much-branched, densely strigillose. Rosette leaves 4–8 cm long, 0.7–1 cm wide, narrowly oblanceolate, remotely shallowly dentate, apex acute, base gradually narrowed to the petiole; cauline leaves 1–7 cm long, 0.3–1.5 cm wide, narrowly oblong to narrowly elliptic or narrowly obovate, remotely shallowly dentate to subentire, apex acute, base narrowly cuneate, short-petiolate to sessile; bracts 1–2.5 cm long, 0.3–1 cm wide, narrowly oblong to narrowly elliptic or narrowly oblanceolate, remotely dentate to entire, apex acute, base narrowly cuneate, short-petiolate to sessile; leaves and bracts grayish green, densely strigillose. Inflorescence lax, usually with lateral branches. One flower per spike opening each day near sunset. Floral tube 1.5–3.5 cm long, ca. 1.5 mm in diameter, densely strigillose to villous and scattered glandular-puberulent. Mature buds 2.5–4 mm in diameter, oblong to ovoid. Sepals 0.3–1.1 cm long, green, pubescence same as floral tube, sepal tips 0.5–2 mm long, erect and appressed or slightly divergent in bud, strigillose. Petals 0.45–1.6 cm long, 0.5–1.7 cm wide, yellow to pale yellow, very broadly obovate, emarginate to truncate. Filaments 4–11 mm long; anthers 2–5.5 mm long, shedding pollen directly on stigma at anthesis, pollen ca. 50% fertile. Ovary 1–1.9 cm long, ca. 1.5 mm in diameter, densely strigillose, sometimes also glandular-puberulent in upper part. Style 2.3–4.5 cm long, the exserted part 8–11 mm long; stigma surrounded by the anthers at anthesis, lobes 3–5 mm long. Capsule 1.5–4.5 cm long, 2–3 mm in diameter, cylindrical, strigillose. Seeds 1.2–1.8 mm long, 0.5–0.9 mm in diameter, ellipsoid to broadly ellipsoid, dark brown, sometimes darker flecked, the surface inconspicuously pitted, Autogamous, permanent structural heterozygote. Chromosome number: n = 7 (⊙14* chromosomes at meiotic metaphase I). Fig. 10.

Phenology. Flowering from April to November.

Distribution (Fig. 11). Dunes and open sandy places along the Atlantic coast of New Jersey, Pennsylvania, Maryland, Virginia, North and South Carolina, Georgia, Florida, Alabama, Mississippi, and Louisiana; Bermuda and Cuba. The inland collection from Iredell Co., North Carolina, presumably represents an introduction.

Specimens from Cultivated Plants. U.S.A. Florida: Pinelas Co., Mullet Key, Fort de Soto Park, 1973, *Long s.n.* [cult. no. 79–41*] (DUSS, M, MO). Santa Rosa Co., Isle of Santa Rosa, 1977, *Binding s.n.* [cult. no. 79–93*] (DUSS, M, MO).—Georgia: Chatham Co., Tybee Island, 1974, *Thorne s.n.* [cult. no. 75–1470*] (DUSS, M, MO).—Maryland: Worcester Co., Ocean City, *Cleland 479* [cult. no. 75–1452*] (DUSS, M, MO).—North Carolina: Cartered Co., Fort Macon, 1968, *Cleland s.n.* [cult. no. 77–555*] (DUSS, M, MO). Dare Co., Kitty Hawk, *Cleland 503* (1952, *Stoutamire s.n.*) [cult. no. 79–49*] (DUSS, M, MO). New Hanover Co., Wrightsville near Wilmington, 1950, *Cleland s.n.* [cult. no. 76–736*] (DUSS, M, MO). Onslow Co., Bear Island, Hammocks State Beach Park, 1973, *Bolick & Flint s.n.* [cult. no. 75–1454*, 75–1462*, 75–1466*] (DUSS, M, MO).—Virginia: City of Virginia Beach: Virginia Beach, 1974, *Straley s.n.* [cult. no. 75–1470a*] (DUSS, M, MO), 1978, *Ellstrand s.n.* [cult. no. 79–92*] (DUSS, M, MO).

Representative Specimens. Bermuda: Southhampton, *Brown 720* (C, GH, PHIL); Cartle Point, *Brown & Britton 975* (PH); Warwick camp, 1921, *Degener s.n.* (BH, MASS), *Hunter 114* (BM); St. George, *Manuel 168* (GH).—Cuba: Habana Prov., Boca de Guanabo, *León 11649* (GH).—U.S.A. Alabama: Baldwin Co., Fort Morgan, *Tracy 7601* (CU, E, F, G, GH, MIN, MO, MSC, NY, PH). Mobile Co., Dauphin Island, *Deramus D464* (MO, NCU).—Delaware: Kent Co., Big Stone Beach, *Larsen 929* (GA, MIN, PENN). Sussex Co., Slaughter Beach, *Larsen 1081* (GA, PENN, PH).—Florida: Bay Co., Panama City, *Hood 3378* (FLAS). Brevard Co., Eau Gallie, *Curtiss 5720* (CONN, FLAS, G, GH, HBG, K, MO, MSC, SIU, US, VT). Broward Co., Fort Lauderdale, 1942, *Knott s.n.* (FLAS). Charlotte Co., leading to Boca Grande bridge, *D'Arcy 1459* (FLAS). Collier Co., Marco Island, *Lakela 29847* (NCU, NY). Dade Co., Miami, *Small & Small 6941* (DUKE, FLAS, G, NCU, PENN, TENN, WVA). Duval Co., Mayport, *Godfrey & Lindsey 56817* (AUA, FSU, GH, IA, USF). Escambia Co., Pensacola, *Brinker 420* (NA). Flagler Co., Flagler Beach, 1940, *West & Arnold s.n.* (FLAS). Franklin Co., Apalachicola, 1897, *Chapman s.n.* (MIN, MO). Gulf Co., Port St. Joe, *Sargent 6169* (KANU, POM, SMU, WIS). Hillsborough Co., Beacon Key, *Todd 281* (USF). Indian River Co., Indian River Island, *D'Arcy 2919* (FLAS). Lee Co., Punta Rassa, *Eaton 1298* (GH, NY, TEX). Manatee Co., Bradenton, Longboat Key, *Long & Lakela 27578* (DS, FLAS, USF). Martin Co., Hobe Sound, *McCart 11215* (FLAS). Nassau Co., Amelia Island, *Hood 4407* (FLAS). Okaloosa Co., Deerland, *Ford 3875* (MO). Okeechobee Co., Okeechobee, 1962, *Craighead s.n.* (FLAS). Orange Co., Orlando, *Gale 137* (FLAS). Palm Beach Co., Juno Beach area, *Lakela 24957* (FLAS, GH, USF). Pinellas Co., St. Petersburg, *Deam 5005* (IND, MIN, NMC). St. Johns Co., St. Augustine, *Bright 3540* (CM). Santa Rosa Co., Santa Rosa Island, *Boufford et al. 18579* (NCU). St. Lucie Co., Fort Pierce, *Burgess 418* (F, NY). Sarasota Co., Venice, *Moldenke 1038* (B, DUKE, MO, MN). Volusia Co., New Smyrna Beach, *Evans et al. 45528* (NY, TENN). Wakulla Co., Mashes Island, *Godfrey 53212* (DUKE, GH, NCU, USF, VDB). Walton Co., Santa Rosa, *Demaree 50306* (NCU, VDB).—Georgia: Chatham Co., Tybee Island, 1918, *Small s.n.* (NY). Glynn Co., Jekyll Island, *Bozeman 6411* (NCU). McIntosh Co., Sapelo Island, *Duncan 20245* (AUA, CAS, DUKE, GH, MISS, SM, TEX, US, USF, WIS). Ware Co., Waycross 1946, *Stifler s.n.* (NCU).—Louisiana: Cameron Pa., Grand Chenier, 1958, *Chamberlain s.n.* (FSU). St. Tammany Pa., Pearlington, *Thieret 29168* (DS, GA, LSU).—Maryland: Worcester Co., Assateague Island, *Windler et al. 3676* (B, CONN, FLAS, FSU, ILLS, KANU, KRAM, MASS, MSC,

NCU, OKLA, PAM).—MISSISSIPPI: Harrison Co., Ship Island, Biloxi, *Demaree 30578* (KANU, MISS, MSC, UARK, VDB). Jackson Co., Horn Island, Ocean Springs, *Demaree 32415* (DUKE, IND, OKLA, TEX).—NEW JERSEY: Cape May Co., Cape May Point, *Mackenzie 4455* (GH, NY).—NORTH CAROLINA: Brunswick Co., Long Beach, *Bradley & Stevenson 3315* (ASC, AUA, B, C, CM, COLO, E, FSU, GA, ILLS, IND, KANU, KE, KRAM, KY, MASS, MIN, MISS, NCU, OKLA, PAM, RSA, SIU, SMS, SMU, TENN, TEX, UARK, USF, VPI, WIS, WVA), *Leonard et al. 1761* (B). Carteret Co., Shackleford Point, 1898, *Ashe s.n.* (FSU, NCU).—VIRGINIA: Accomac Co., Parramore Island, *McVaugh 5738* (GH). City of Virginia Beach [as Princess Anne Co.]: 1893: *Heller s.n.* (P), *Fernald & Long 4077* (GH).

Oenothera humifusa is very similar to *O. drummondii* in growth form and habitat requirements. It is an autogomous, permanent structural heterozygotic species apparently directly derived from *O. drummondii.* The distribution of *O. humifusa* along the Atlantic coast extends north to New Jersey and Pennsylvania, whereas that of *O. drummondii* extends only to North Carolina. Their ranges overlap along the coast between North Carolina and Louisiana. Common to both *O. humifusa* and *O. drummondii* are the appressed pubescence, the usually procumbent habit, as well as the tendency to grow for several years. Two basic types can be distinguished; however, they are connected by intermediate forms.

Plants of the first form are somewhat procumbent; they have subentire cauline leaves and bracts, and mostly emarginate petals. The lectotype of *O. humifusa* also corresponds with this form to which the following cultivated strains belong: Florida, Pinellas Co., DUSS cult. no. 79–41; Florida, Santa Rosa, 79–93; Georgia, Thatham Co., 75–1470; North Carolina, Wrightsville, 76–736. Plants of the second form are more upright with more deeply divided leaves and retuse petals. Examples are: North Carolina, Kitty Hawk, 79–49; Virginia, Virginia Beach, 75–1470a & 79–92. An intermediate is: North Carolina, Fort Macon, 77–555.

Experimental crosses between *O. humifusa* and other species showed that both complexes of *O. humifusa* are transmitted by the egg cell, but only one complex is transmitted by the pollen. For example, crosses between *O. humifusa* (Onslow Co.) and *O. laciniata* (Padre Island) or *O. laciniata* (Mississippi) resulted in F_1 offspring of two phenotypes (DUSS cult. no. 76–438, 78–271, 79–187 & 79–189). In contrast, crosses between *O. drummondii* subsp. *drummondii* (Bermuda) or *O. drummondii* subsp. *thalassaphila* (Cabo San Lucas) and *O. humifusa* produced only one phenotye (76–437 & 76–451). These crosses also showed that although the two genome complexes of *O. humifusa* were slightly different morphologically, both apparently were derived from *O. drummondii.*

Oenothera section **Oenothera** subsection **Nutantigemma** W. Dietrich & W. L. Wagner, Ann. Missouri Bot. Gard. 74: 145. 1987.—TYPE: *Oenothera pubescens* Willdenow ex Sprengel.

Erect or ascending annual or probably biennial herbs, forming rosettes; stems green but often flushed with red, simple or with a branched main stem and arcuating branches arising from the rosette, strigillose or strigillose and villous, rarely also glandular-puberulent in upper parts. Rosette leaves narrowly oblanceolate, deeply parted to subentire, apex acute, gradually narrowed into the petiole; cauline leaves very narrowly elliptic to elliptic, narrowly oblanceolate to oblanceolate, narrowly lanceolate to lanceolate or narrowly oblong, deeply parted to subentire, apex acute, base narrowly cuneate to subcordate; all leaves strigillose or villous, rarely in *O. pubescens* also glandular-puberulent. Inflorescence simple or with lateral branches, the apex often nodding. Young buds erect, the older ones with a recurved floral tube and nodding, becoming erect just before opening. Usually one flower per spike opening each day near sunset. Petals yellow, changing to red or orange when wilted, broadly ovate to very broadly ovate, rarely suborbicular, truncate to retuse. Stigma elevated above the anthers at anthesis or surrounded by the anthers and pollen shed directly on the stigma. Capsules cylindrical. Seeds ellipsoid to subglobose, brown to dark brown, often darker flecked, 0.9–1.6 mm long. Self-compatible. Chromosome number: n = 7.

The principal reason for according these three species (*O. pubescens*-group of Stubbe and Raven, 1979) the status of a subsection is the sterility of hybrids in crosses between members of subsections *Raimannia* and *Nutantigemma.* The nodding flower buds of species of subsect. *Nutantigemma* clearly differentiate them from the species of subsect. *Munzia* and subsect. *Raimannia.* This feature is found also in the white-flowered sections *Kleinia, Anogra* (Spach) Endlicher, *Ravenia,* occasionally in *O. caespitosa* subsp. *navajoensis* W. L. Wagner, Stockhouse & Klein (sect. *Pachylophus*) and in the yellow-flowered sect. *Eremia.* In *O. perennis* Linnaeus and sometimes *O. fruticosa* Linnaeus (both sect. *Kneiffia*), which have yellow flowers, the shoot apices, but not the flower buds, are bent downwards; a similar condition occurs in *O. speciosa* Nuttall [sect. *Xylopleurum* (Spach) Endlicher]. Nodding buds appear to represent, at least for the most part, a plesiomorphic feature in the common ancestors of *Oenothera* rather than one of convergence. Another important argument for considering this group a subsection is that these three species have a geographical distribution completely allopatric from that of subsect. *Raimannia.* The species of subsect. *Nutantigemma* grow exclusively in montane habitats, approximately 1500 m or higher, whereas the species of subsect. *Raimannia* grow only at lower elevations. *Oenothera breedlovei* is a self-compatible, bivalent-forming species; *O. pubescens* is a permanent structural heterozygote; *O. tamrae,* not yet in cultivation and known only from the type collection, is probably an outcrossing bivalent former.

12. Oenothera breedlovei W. Dietrich & W. L. Wagner, Ann. Missouri Bot. Gard. 74: 146. 1987.—Type: Mexico. Baja California Sur: granitic slopes surrounding long interior valley [La Laguna], S of Pico La Aguja, Sierra La Laguna, 6300–6700 ft, *Breedlove 43362* (holotype: MO-2696034!).

Erect annual or probably biennial herbs, forming rosettes; stems 2–3 (–5) dm long, simple or with a branched main stem and arcuating lateral branches arising from the rosette, usually flushed with red, densely strigillose, sometimes also scattered villous. Rosette leaves 4–12 cm long, 0.5–2 cm wide, narrowly oblanceolate, pinnately parted to remotely and shallowly dentate, apex acute, gradually narrowed to the petiole; cauline leaves 2–5 cm long, 0.5–2 cm wide, very narrowly elliptic to lanceolate, usually pinnately parted, apex acute, base narrowly cuneate, short-petiolate to sessile; bracts 1.5–3 cm long, 0.5–1.5 cm wide, lanceolate to narrowly ovate, deeply parted to remotely shallowly dentate, apex acute, base broadly cuneate to narrowly cuneate; leaves and bracts densely strigillose and also sparsely villous. Inflorescence usually with lateral branches, young buds erect, the older ones nodding, becoming erect just before opening. Usually one flower per spike opening each day near sunset. Floral tube 2.4–4 cm long, ca. 1 mm in diameter, flushed with red, densely to sparsely strigillose, villous, and glandular-puberulent. Mature buds 3–5 mm in diameter at the base, oblong. Sepals 1.2–2.2 cm long, yellowish, often flushed with red, also red-maculate and striped at the margins, densely to scattered glandular-puberulent, sepal tips 0.5–1 mm long, erect and appressed in bud, strigillose to villous. Petals 1.6–3.5 cm long, 1.6–3.7 cm wide, yellow, broadly obovate to very broadly obovate, retuse. Filaments 1–1.2 cm long; anthers 4–12 mm long, pollen ca. 90–100% fertile. Ovary 1–2.5 cm long, ca. 1.5 mm in diameter, densely strigillose. Style 4–6.5 cm long, the exserted part 1.7–2.5 cm long; stigma elevated above the anthers at anthesis, lobes 3–6 mm long. Capsule 1.8–4.6 cm long, 3–3.5 mm in diameter, cylindrical, densely strigillose. Seeds 1.2–1.6 mm long, 0.6–0.7 mm in diameter, ellipsoid to broadly ellipsoid, brown to dark brown, often with darker flecks, the surface pitted. Self-compatible, but modally outcrossing. Chromosome number: n = 7 (7_{II}* at meiotic metaphase I). Fig. 13.

Phenology. Flowering from August to November.

Distribution (Fig. 12). Laguna Mountains, Baja California Sur, Mexico.

SPECIMENS FROM CULTIVATED PLANTS. MEXICO. BAJA CALIFORNIA SUR: Laguna Mts, *Breedlove & Axelrod 43231, 43340, 43362* [cult. no. 78–249*, 78–250*, 78–251*] (DUSS, M, MO).

ADDITIONAL SPECIMENS EXAMINED. MEXICO. BAJA CALIFORNIA SUR: Sierra de La Laguna (23°34′N, 110°00′W), 1890, *Brandegee s.n.* (UC); La Chuparosa, 1899, *Brandegee s.n.* (UC); Sierra de San Francisquito, *Brandegee 220* (UC); La Laguna, Sierra de La Victoria, *Chambers 892* (DS, MEX, SD, UC, US); La Luna near Todos Santos, Chuparosa and Francisquito (23°29–31′N, 109°47–55′W), *Cody s.n.* (MO); Sierra de La Laguna, 1941, *Hammerly s.n.* (DS, GH); La Laguna, Sierra de La Laguna (23°32′N, 109°30′W), *Jones 27044* (DS, NY, POM, UC, US); Laguna, Sierra de La Laguna, *Jones 27206* (MO); Valle de La Laguna, 1944, *Martínez s.n.* (MEXU); Sierra La Laguna, S Pico La Aguja, 1900–2050 m, *Breedlove & Axelrod 43231, 43340, 43362* (MO).

Oenothera breedlovei is a large-flowered, bivalent-forming species (7_{II} at meiotic metaphase I), and thus appears to be a generalized member of subsect. *Nutantigemma.* It also has a narrow and apparently relictual distribution in the Laguna Mountains of Baja California Sur. All plants of *O. breedlovei* examined by self-pollination were self-compatible.

Because of the extensive morphological variation in *O. pubescens,* it is rather difficult to distinguish all of its populations from *O. breedlovei,* although most are quite distinct. The most reliable visible character of *O. breedlovei* is that the stigma is well elevated above the anthers, which undoubtedly greatly increases the outcrossing rate. The petals of *O. breedlovei* are 2.5–3.5 cm long, whereas those of *O. pubescens* are usually only 0.5–2.5 cm long. However, rare forms of *O. pubescens,* which occur in scattered localities in Mexico, have petals that are nearly as long as those of *O. breedlovei* (e.g., Sierra Torticillas, Michoacán, *Hinton 15014;* Otinapa, Durango, *Maysilles 7326;* Tlaltenango, Zacatecas, *Straw 1808*), but they have flowers with the stigma surrounded by the anthers and pollen ca. 50–60% fertile. The rosette leaves and the lower stem leaves of *O. breedlovei* are deeply divided nearly to the midrib and have a large terminal lobe, whereas those of *O. pubescens* are less deeply divided or merely dentate. Unfortunately, this characteristic can rarely be used in making a determination of herbarium material, since the rosette leaves and lower stem leaves of the plants are absent in most specimens, which are usually collected in a flowering state. The pubescence of the upper leaves of *O. breedlovei* is always clearly appressed, which gives the plants a grayish color, even in the

FIG. 13. *Oenothera breedlovei* and *O. pubescens*. *O. breedlovei* (*Breedlove & Axelrod 43231;* cult. no. 78–249; MO): a. Inflorescence. *O. pubescens* (*Martin 510,* Mexico, Oaxaca, Dist. Ixtlán; MO): b. Inflorescence.

inflorescence; glandular pubescence has never been observed in this species. In contrast, *O. pubescens* is often only sparsely pubescent, and glandular pubescence is usually evident in the upper part of the expanding shoot. These species can be distinguished reliably, of course, by their pollen fertility, which is ca. 90–100% in *O. breedlovei* and only up to ca. 50–70% in *O. pubescens.*

13. Oenothera tamrae W. Dietrich & W. L. Wagner, Ann. Missouri Bot. Gard. 74: 146. 1987.—TYPE: MEXICO. Nayarit: Sierra Madre, near Santa Teresa, territory of Tepic, 8 Aug 1897, *Rose 2133* (holotype: US-301038!; isotypes: NY! UC!). Fig. 14.

Erect annual or biennial herbs, probably forming rosettes; stems 2–4 dm long, simple or with oblique ascending lateral branches arising from the rosettes, densely strigillose and densely to sparsely villous. Cauline leaves 4–8 cm long, 0.8–1.8 cm wide, narrowly elliptic or narrowly lanceolate to lanceolate, pinnately parted or remotely and bluntly dentate to subentire, apex acute, base narrowly cuneate, sessile; bracts 3–4.5 cm long, 0.7–1.5 cm wide, narrowly lanceolate to lanceolate, remotely and bluntly dentate, apex acute, base narrowly cuneate, sessile; leaves and bracts strigillose. Inflorescence simple or with lateral branches, nodding. Usually one Flower per spike opening each day near sunset. Floral tube 3.5–4.2 cm long, 1.5–2 cm in diameter, flushed with red, sparsely villous and sparsely glandular-puberulent. Mature buds 5–7 mm in diameter at the base, cylindrical to narrowly ovoid, nodding before anthesis. Sepals 1.8–2.5 cm long, yellowish, often flushed with red and striped red at the margins, pubescence like floral tube, sepal tips ca. 0.5 mm long, erect in bud, strigillose. Petals 2–3.5 cm long, 3–4 cm wide, yellow, very broadly obovate, retuse. Filaments 1.5–1.7 cm long; anthers 6–9 mm long, pollen ca. 90–100% fertile. Ovary 1.8–2.6 cm long, ca. 2 mm in diameter, densely strigillose and densely villous, apex also glandular-puberulent. Style 5.3–6.4 cm long, the exserted part 1.8–2.2 cm long; stigma elevated above the anthers at anthesis, lobes 5–9 mm long. Capsule 4–4.5 cm long, 3–4 mm in diameter, cylindrical, pubescence like that of ovary, but less dense. Seeds 1–1.1 mm long, ca. 0.7 mm in diameter, broadly ellipsoid, brown with dark red flecks, the surface pitted. Chromosome number unknown. Fig. 14.

Distribution (Fig. 12). Known only from the type locality, near Santa Teresa in the Sierra Madre Occidental, Nayarit, Mexico.

Our description of *O. tamrae* is based entirely on the type collection made by J. N. Rose in 1897. D. E. Breedlove searched for *O. tamrae* in October 1979 to obtain seeds for experimental cultivation but unfortunately was unsuccessful. His search for areas near the type locality yielded only a new locality of *O. pubescens* (*Breedlove 44465*).

The three sheets of the type collection clearly show that they represent a distinct species related to *O. pubescens* and *O. breedlovei.* Because *O. tamrae* has the stigma elevated above the anthers as well as a pollen fertility (stainable pollen) of 91% (NY specimen), we believe that, unlike *O. pubescens,* it is a bivalent-forming species. Such a high percentage of pollen fertility is not known in any permanent structural heterozygote of *Oenothera,* and the pollen fertility of bivalent-forming species of *Oenothera* is always 90–100%. Representative values for the percentage of stainable pollen in *O. pubescens* are as follows: 1977, *Rehwinkel s.n.,* Mexico, Chiapas, 52%; *Breedlove 44465,* Mexico, Nayarit, 62%; *Lugo 509,* Ecuador, Chimborazo, 58%; *Santarius 2155,* Peru, Junín, 60%. Between 400 and 600 pollen grains were counted for each determination.

The size of flowers is also an indication, although a weak one, for our assumption that *O. tamrae* is a bivalent-forming species. Large flowers occur only infrequently in *O. pubescens,* and in the vicinity of Santa Teresa the petals of plants of *O. pubescens* studied by Breedlove were at most 2 cm long, which was considerably smaller than the petals of the type of *O. tamrae,* which are 2–3.5 cm long.

The five plants of the type collection of *O. tamrae* vary only slightly in leaf shape. The specimen at NY has sinuate to dentate stem leaves, whereas the two specimens at US have only weakly dentate leaves; the leaf dentation of the right-

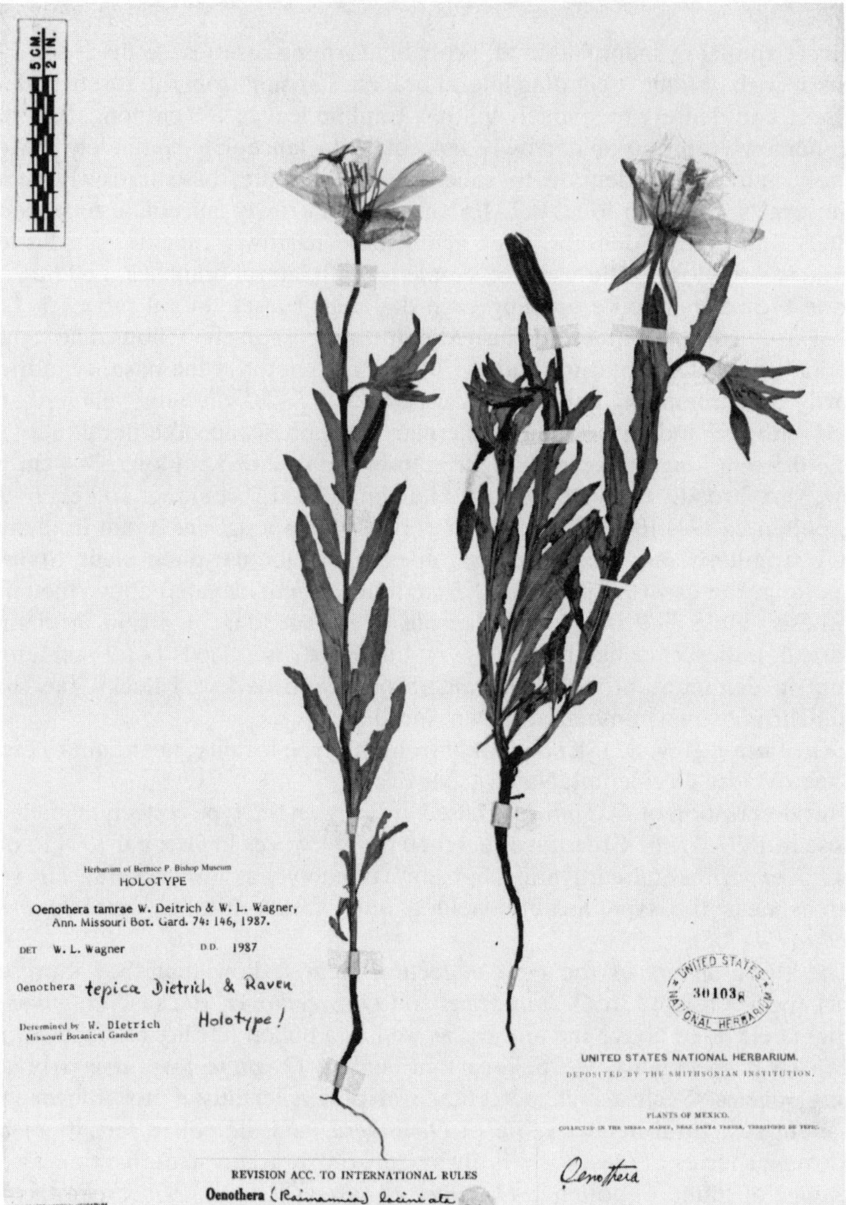

FIG. 14. Photograph of holotype of *Oenothera tamrae* (US-301038).

hand plant mounted on the UC herbarium sheet is intermediate, whereas the leaves on the left-hand one correspond with those of the US specimen.

14. Oenothera pubescens Willdenow ex Sprengel, Syst. veg. 2: 229. 1825. *Oenothera laciniata* var. *pubescens* (Willdenow ex Sprengel) Munz, Amer. J. Bot. 22: 656. 1935. *Oenothera laciniata* subsp. *pubescens* (Willdenow ex Sprengel)

Munz, Fl. N. Amer. II. 5: 109. 1965.—TYPE: ECUADOR. *Humboldt & Bonpland s.n., Herb. Willdenow 7177* (holotype: B!, photograph: F-14008!).

Oenothera stuebelii Hieronymus, Bot. Jahrb. Syst. 21: 327. 1895.—TYPE: ECUADOR. Imbabura: Loma and vicinity of La Canaballa, 2000–2300 m, 1 Jul 1871, *Stübel 161b* (holotype: B, destroyed, photographs: BH! F! POM! UC! with fragments).

Oenothera nyctaginiifolia Small, Bull. Torrey Bot. Club 27: 278. 1900.—TYPE: U.S.A. Arizona: Coconino Co., Flagstaff, 5 Sep 1894, *Toumey s.n.* (holotype: NY!; isotype: UC!).

Raimannia colimae Rose ex Sprague & Riley, Bull. Misc. Inform. 1921: 200. 1921.—TYPE: MEXICO. Jalisco: Volcano of Colima, 14 Jul 1892, *Jones 232* (holotype: US-237424!; isotypes: MO! POM!).

Raimannia confusa Rose ex Sprague & Riley, Bull. Misc. Inform. 1921: 200. 1921.—TYPE: MEXICO. Hidalgo: Sierra de Pachuca, 21 Jul 1901, *Rose & Hay 5636* (holotype: US-395404!; isotype: GH!).

Anogra amplexicaulis Wooton & Standley, Contr. U.S. Natl. Herb. 16(4): 150. 1913. *Oenothera amplexicaulis* (Wooton & Standley) Tidestrom in Tidestrom & Kittel, Fl. Ariz. & N. Mex. 274. 1941.—TYPE: U.S.A. New Mexico: Grant Co., on a sandbar along the Mimbres River, ca. 1830 m, 1 Jul 1904, *Metcalfe 1054* (holotype: US-497937!; isotypes: ARIZ! CAS! ISC! LL! MIN! MO! NY! POM! NMC! UNM!).

Oenothera pennellii Munz, Leafl. West. Bot. 2(9): 156. 1939.—TYPE: MEXICO. Nuevo León: Sierra Madre Oriental, from grassy slope, Mt. "El Infiernilla," Pablillo, SE Galeana, 29 Jun 1934, *Pennell 17139* (holotype: US-1640419!; isotypes: PH! POM!).

Erect annual or biennial herbs, forming rosettes; stems 0.5–5 (–8) dm long, simple or with a branched main stem and arcuate to procumbent lateral branches arising from the rosette, densely to sparsely strigillose, sometimes in upper part also villous and glandular-puberulent. Rosette leaves 5–14 cm long, 0.5–2.5 cm wide, narrowly oblanceolate, pinnately parted to subentire, apex acute, gradually narrowed into the petiole; cauline leaves 2–8 cm long, 0.5–2.5 cm wide, narrowly oblanceolate, narrowly oblong to lanceolate or very narrowly elliptic to elliptic, pinnately parted to subentire, apex acute, base narrowly cuneate to subcordate, short-petiolate to sessile; bracts 1.5–6 cm long, 0.5–2.5 cm wide, narrowly oblong, narrowly elliptic, narrowly lanceolate to lanceolate or narrowly ovate, dentate to subentire, apex acute, base narrowly cuneate to truncate, short-petiolate to sessile; leaves and bracts densely to sparsely strigillose, the upper ones sometimes also villous and glandular-puberulent. Inflorescence simple or with lateral branches, the apex often curved. Usually one flower per spike opening each day near sunset. Floral tube 1.5–5 cm long, villous and glandular-puberulent, sometimes also sparsely strigillose. Mature buds 2.5–5 mm in diameter at base, nodding but becoming erect just before anthesis, narrowly cylindrical to lanceoloid. Sepals 0.5–2.5 cm long, yellowish, often flushed with red and striped red at the margins, often dark red-flecked, pubescence like that of floral tube, sepal tips 0.1–1 mm long, appressed in bud, strigillose. Petals 0.5–2.5 (–3.5) cm long, 0.6–2.6 (–3.5) cm wide, yellow, aging red, broadly obovate to very broadly obovate, sometimes rotund, apex rounded to truncate, rarely subacute, rarely with a small auricle on each side at base. Filaments 6–18 mm long; anthers (2–) 3–9 mm long, shedding pollen directly on the stigma at anthesis, pollen ca. 50 (–70)% fertile. Ovary 1–2 cm long, ca. 1.5 mm in diameter, densely strigillose, often also villous and glandular-puberulent. Style 2–6 cm long, the visible part 0.5–2 cm long; stigma surrounded by the anthers at anthesis, lobes 2–7 mm long. Capsule 2–4.5 cm long, 2–4 mm in diameter, cylindrical pubescence like that of ovary. Seeds 0.9–1.5 mm long, 0.6–1 mm in diameter, broadly ellipsoid to subglobose, brown, sometimes dark-flecked, the surface pitted. Autogamous, permanent structural heterozygote. Chromosome number: n = 7 (⊙14*, or ⊙12 and 1_{II}** at meiotic metaphase I). Fig. 13.

Phenology. Flowering throughout the year, or in April to August in the northern part of the range.

Distribution (Fig. 12). Scattered to locally common in open sites in montane areas, 1500–3900 m, from New Mexico to Arizona, one locality in southeastern California, Mexico to Guatemala, and in South America in the Andes of Colombia, Ecuador, and Peru south to the province of Junín.

Specimens from Cultivated Plants. Ecuador. Chimborazo: Cerro Chiguazo, 2900 m, *Lugo 509* [cult. no. 77–624*] (DUSS, M, MO).—Pichincha: Guaillabama, 1972, *Müller s.n.* [cult. no. 80–550*] (DUSS, M, MO).—Mexico. Chiapas: San Lucas Zapotal, 2400 m, *Breedlove 37294* [cult. no. 77–618*] (DUSS, M, MO); San Cristóbal de las Casas, 1977, *Rehwinkel s.n.* [cult. no. 78–073*, 78–277*] (DUSS, M, MO).—Coahuila: Sierra del Carmen, Pico de Centinela, 2000 m, *Henrickson 11643* [cult, no. 78–254*] (DUSS, M, MO).—Cuernavaca: Cuernavaca, 1974, *González-Medrano s.n.* [cult. no. 78–252*] (DUSS, M, MO).—Durango: 5 km W of Navios, 2700 m, *Breedlove 36502* [cult. no. 75–1512*] (DUSS, MO).—México: Amecameca, *Breedlove 36519* [cult. no. 75–1514*] (DUSS, M, MO); San Cristóbal Ecatepec, *Rzedowski s.n.* [cult. no. 77–619*] (DUSS, MO, MO).—Jalisco: Los Volcanes, *Breedlove 35872* [cult. no. 77–622*] (DUSS, M, MO).—Nayarit: Arroyo Santa Rosa W of Santa Teresa, 100 km NE of Tepic, 2095 m, *Breedlove 44465* [cult. no. 80–305a*] (DUSS, M, MO), *Breedlove 59174A* [cult. no. 84–263*] (MO).—Querétaro: Amealco, 2700 m, *Breedlove 36512* [cult. no. 77–616a*] (DUSS, M, MO).—Veracruz: Las Vegas, 40 km N of Xalapa, 1974, *Dorantes & Acosta s.n.* [cult. no. 77–620*] (DUSS, M, MO).—Peru: Lima: Huarochiri, between Surco and Puente Quitasombrero, 1974, *Encarnación s.n.* [cult. no. 76–787*] (DUSS, M, MO); between Matucana and Surco, *Santarius 2326* [cult. no. 69-2326*] (DUSS).—Junín: Río Mantaro, 1 km S of Pachacayo, *Santarius 2190, 2198* [cult. no. 69–2190*, 2198*] (DUSS, MO).—Ayacucho: W of Ayacucho, *Santarius 2235* [cult. no. 69–2235*] (DUSS, MO), *Santarius 2238* [cult. no. 69–2238**] (DUSS).—Spain. Bot. Garden Madrid, 1806 (MA).

Representative Specimens [material from South America cited in Dietrich (1977)]. Galapagos Islands. Charles Island: Post Office Bay, *Howell 8821* (CAS).—Guatemala. Chimaltenango: Barranca de la Sierra, SE of Patzun (ca. 14°41′N, 91°01′W), *Standley 61493* (F).—Guatemala: San Juan Sacatepéquez (14°43′N, 90°39′W), *Standley 59262* (F).—Huehuetenango: Wahshaklahung on San Mateo Ixtatán (15°50′N, 91°29′W), *Breedlove 11563* (MO); Huehuetenango (15°20′N, 91°28′W), *Standley 65762* (F).—Quezaltenango: 3 mi E of Quezaltenango along road to San Cristóbal Totonicapán (14°50′N, 91°31′W), *Breedlove 11449* (MO).—Sacatepéquez: San Mateo del Milpar, 10 km from Antigua (near 14°34′N, 90°44′W), *Molina 15491* (F, GH).—San Marcos: vicinity of Sibinal (15°08′N, 92°03′W), *Steyermark 35948* (F); Cerro Tumbador, Sierra Madre Mts, 15 km W of San Marcos (near 14°58′N, 91°48′W), *Williams et al. 23098* (F).—Mexico. Aguascalientes: road to Calvillo 21 mi W of Aguascalientes (near 21°51′N, 102°43′W), *Koelz 119* (MICH).—Chiapas: Tenejapa (16°49′N, 92°31′W), *Breedlove 10731* (MO); Chiapas (16°42′N, 93°00′W), *Ghiesbreght 677* (CONN, GH, MO, NY, YU).—Chihuahua: Majalca (28°52′N, 106°20′W), *LeSueur 804* (F, IND, TEX, UC, US); Chuhuichupa (29°38′N, 108°22′W), *LeSueur 807* (F, TEX); Cima (28°23′N, 107°04′W), *LeSueur 809* (CAS, F, GH, MO, TEX); near Colonia Juárez (near 30°19′N, 108°05′W), *Nelson 6112* (NY, US); Cerro Mohinora, 10 mi SW of Guadalupe y Calvo (26°06′N, 107°04′W), *Straw & Forman 1918* (MICH, RSA); 5 mi E of Ciudad Guerrero on Hwy 16 (ca. 28°33′N, 107°30′W), *Stuessy 1057* (F, TEX).—Coahuila: Sierra del Carmen, Cañon de Centinela, southside of Pico de Centinela (29°03′N, 102°35′W), *Henrickson 11643* (MO); Sierra de Parras (25°24′N, 102°24′W), *Purpus 4624* (UC).—Durango: 11 km SW of La Ciudad near Buenos Aires (ca. 24°23′N, 106°03′W), *Breedlove 36482* (MO); 6 km SW of Guadalupe Victoria along Hwy 40,

Henrickson 1748 (DS, DUKE, MICH, RSA); 3 mi NE of Otinapa (24° 11′ N, 105° 02′ W), *Maysilles 7326* (MICH, RSA); Tepehuanes (25° 21′ N, 105° 44′W), *Palmer, E. 40* (GH, MO, NY, UC, US); vicinity of Durango (24°02′N, 104°40′W), *Palmer, E. 155* (C, F, GH, MO, UC, US); 5.7 mi W of El Salto on Hwy 40, *Wagner & Solomon 4276* (MO); 19 mi SW of Durango, *Waterfall & Wallis 13462* (ISC, OKL, RSA, SMU).—Guanajuato: San Miguel Allende (21°00′N, 100°30′W), *Coon 089* (GH); Rancho Las Cruces, Salvatierra (20°13′N, 100°53′W), *Flores 64* (ENCB); 30 mi E of San Luis de la Paz toward Xichú (near 21°25′N, 100°06′W), *Straw & Forman 1469* (RSA).—Guerrero: vicinity of Omiltemi, ca. 60 km W of Chilpancingo (ca. 17°30′N, 99°40′W), *Fillet 637–164* (CS, DS, RSA, US).—Hidalgo: ca. 25 km N of Pachuca on Hwy 105 (near 20°07′N, 98°44′W), *Conrad 3312, 3313* (MO); Metepec Station (19°57′N, 98°39′W), *Pringle 10314* (C, CU, E, F, G, GH, HBG, MEXU, MIN, MO, MSC, NY, P, PH, US, VT); Cerro Xihuingo, Ciudad Tepeapulco (19°47′N, 98°33′W), *Rzedowski 31445* (ENCB).—Jalisco: El Terrero on Hwy 110, Km 59–60, 20 mi WSW of Jiquilpan (19°45′N, 102°45′W), *Gregory & Eiten 102* (MICH); Nevado de Colima (Nevado de Zapotlán), a few mi S of Ciudad Guzmán (Zapotlán) (19°33′N, 103°38′W), *Gregory & Eiten 285* (MICH, MO); SW of Ojuelos near Km 31 on road to Aguascalientes above Presa de Valerio (near 21°11′N, 102°43′W), *McVaugh 16882* (MICH, RSA, TEX, US); barranca al W of Atenquique, Ciudad Tuxpán (19°33′N, 103°24′W), *Rzedowski 21907* (ENCB); E of Mamantlán, ca. 15 mi SSE of Autlán by way of Chante (near 19°36′N, 104°10′W), *Wilbur 1839* (MICH).—Distrito Federal: Km W of San Juan Teotihuacán, Loma de las Calaveras, Hacienda la Cadena (ca. 19°41′N, 98°52′W), *Cruz 704* (DS); ¼ mi W of mountain pass on highway to Temascaltepec, *Dziekanowski et al. 1949* (UMO); Cerro de Sacramento, Amecameca (19°07′N, 98°46′W), *Franco s.n.* (ENCB); ladera E of Cerro de la Tijera, 1 km SE of Sta. Ana, Chalco (19°16′N, 98°54′W), *Pineda 787* (ENCB, MO); Tlayacampa, 2 km SE of Barrientos, Tlalnepantla (19°33′N, 99°12′W), *Pineda 877* (ENCB); 3 km NE of Capula, Sultepec (18°52′N, 99°57′W), *Sancedo 152* (DS, ENCB), *Sancedo 169* (DS, ENCB, F, LL, MICH, MO, NY, UMO, VDB).—México: 5.4 mi N of Morelos State line and 8.7 mi N of Tres Cumbres, *Kral 25247* (MO, VDB); Tlalpan (19°17′N, 99°10′W), *Rose & Painter 6451* (GH, NY, US); Contreras (19°18′N, 99°17′W), *Rzedowski 20935* (DS, ENCB, ISC, MICH, MO, TEX).—Michoacán: 16 mi N of Uruapan, rd. to Parícutin (near 19°25′N, 101°58′W), *Breedlove 15773* (DS); Zitácuaro–La Mesa (19°32′N, 100°24′W), *Hinton 11947* (DS, F, G, GH, ILL, LL, MICH, MIN, MO, NY, P, POM, TEX, U, UC, US); Sierra Torricillas, Coalcomán (18°47′N, 103°09′W), *Hinton 15014* (ARIZ, GH, LL, NA, NY, PH, TEX, US); 1.7 mi W of Tuxpán on Hwy 15 (19°34′N, 100°28′W), *Kral 25190* (ENCB, FLAS, VDB); Tancítaro (19°20′N, 102°22′W), *Leavenworth 244* (ARIZ, F, GH, ILL, MICH, MO, NA); 5.1 km S of Pátzcuaro on road to Opopeo (ca. 19°31′N, 101°36′W), *Ugent & Flores 2185* (DS); SSW of Morelia on road to Villa Madero (just NE of Tiripetio) (19°35′N, 101°19′W), *Ugent & Flores 5798* (DS).—Morelos: 12 mi NE of Cuautla, *Harris 32* (MICH); Cuernavaca (18°55′N, 99°15′W), *Kenoyer 422* (MICH); Tepoztlán (18°59′N, 99°06′W), *Redfield 9* (US).—Nuevo León: El León (26°01′N, 98°53′W), *Hartweg 11* (E); Sierra Madre Oriental, Cienaguillas to Pablillo, 15 mi SW of Galeana (near 24°50′N, 100°04′W), 1934, *Mueller s.n.* (GH, POM, TEX).—Oaxaca: 1.5 mi S of San José del Pacífico, road to Puerto Angel (near 16°10′N, 96°30′W), *Breedlove 15858* (DS); 20 km S of Nochixtlán along Hwy 190 to Oaxaca (near 17°28′N, 97°14′W), *Breedlove 35872* (CAS, MO); Km 218 Hwy 175 at La Cumbre, ca. 30 km NE of Oaxaca (17°11′N, 96°36′W), *Conrad 3076, 3077* (MO); Cuyamecalco, Cuicatlán (17°58′N, 96°50′W), *Conzatti 2430* (US); Oaxaca (17°02′N, 96°44′W), *Cuming 1859* (G); 12 km N of Ixtlán de Juárez on Hwy 175, road to Valle Nacional (near 17°20′N, 96°29′W), *King 2026* (MICH); 10 mi N of Telixtlahuacán (near 17°17′N, 96°54′W), *Pipley 14668* (DS).—Puebla: Puebla (19°03′N, 98°12′W), *Arsène 142* (US); Chinantla (18°12′N, 98°15′W), *Liebmann 3251* (C, GH, UC); few mi W of San Martín Texmelucan toward Río Frío (19°17′N, 98°26′W), *Manning 55104a* (GH); near Huauchinango (near 20°11′N, 98°03′W), *Sharp 45379* (RSA, TENN); SW slopes of Volcán Citlaltepetl (Pico de Orizaba), below Tesmalaquilla (19°01′N, 97°16′W), *Ugent & Flores 2343* (ENCB); near San Hipolito Tochiltenango, Ciudad Tepeaca (18°58′N, 97°54′W), *Weber 338* (ENCB).—Querétaro: 10 mi E of San Juan del Río on Hwy 45 (ca. 20°23′N, 100°00′W), *Breedlove 7212* (MO); 3 km E of Amealco (20°11′N, 100°09′W), *Breedlove 36512* (CAS, MO).—San Luis Potosí: Alvarez (22°03′N, 100°37′W), *Palmer, E. 207* (F, GH, MO, NY, US); vicinity of San Luis Potosí (22°09′N, 100°59′W), *Parry & Palmer, E. 253* (E, GH, MO, NY, P, PH, US, VT).—Sinaloa: Cerro de la Sandía, NE of Panuco (near 23°25′N, 105°55′W), *Pennell 20122* (PH).—Sonora: Sierra de El Tigre, Las Tierritas de El Temblor (30°35′N, 109°12′W), *White 3441* (ARIZ, G, MICH).—Tlaxcala: 3 mi NE of Tlaxco (ca. 19°37′N, 98°07′W), *Johns 588* (MICH, US); 4 km W of Apizaco (19°25′N, 98°09′W), 1967, *Ruiz s.n.* (ENCB).—Veracruz: Jalapa (19°32′N, 96°55′W), *Schiede 529* (NY); Orizaba (18°51′N, 97°06′W), *Seaton 177* (F, GH, US); Las Vigas near La Joya (19°38′N, 97°05′W), *Sharp 45577* (GH,

MEXU, RSA, TENN); Colonia Veinte de Noviembre, Ciudad Perote (19°34′N, 97°15′W), *Ventura 4689* (ARIZ, CAS, ENCB, MICH, TEX, UMO).—Zacatecas: 8 mi N of Sombrerete on Hwy 45 (ca. 23°38′N, 103°39′W), *Breedlove 14340* (MO); 15 air mi NE of Estación Camacho on NW slopes near summit Pico de Teyra (near 24°25′N, 102°18′W), *Henrickson 13455b* (MO); Sierra Madre Oriental, Mt. "El Temerosa," N of Aranzazu (24°38′N, 101°28′W), *Pennell 17466* (PH, US); 5 mi W of Sánchez Román (Tlaltenango) (21°47′N, 103°19′W), *Straw 1808* (RSA).—U.S.A. Arizona: Apache Co., Black River, White Mts, *Gooding 548* (E, G, GH, NY, RSA). Cochise Co., Barefoot Park, Chiricahua Mts, *Blumer 1559* (ARIZ, DS, E, F, KSC, MIN, MO, NMC, NY). Coconino Co., Flagstaff, *Demaree 43779* (ASC). Gila Co., Aztec Peak, *Wagner 358* (DUKE). Graham Co., Mt. Graham, *Kearney & Peebles 9819* (ARIZ, DS, MICH). Greenlee Co., Hwy 666, Rose Peak, *Pinkava et al. 13762* (ASC, NCU, UT). Navajo Co., 10 mi W of McNary, White Mts, *Goodman & Hitchcock 1330* (MO). Pima Co., Rincon Mts, *Blumer 3458* (ARIZ, DS, E, F, GH, ISC, MO, UC). Pinal Co., Sacaton, *Harrison 1861* (ARIZ). Santa Cruz Co., Santa Rita Mts, 1903, *Jones s.n.* (DS, POM). Yavapai Co., Prescott, *Peebles et al. 8873* (ARIZ, F, POM).—California: San Bernardino Co., Newberry Spring, Mohave Valley, 1884, *Lemmon s.n.* (G, GH, UC, US).—New Mexico: Catron Co., Datil Mts, *Fletcher 1271* (UNM). Grant Co., Burro Mts, *Metcalfe 196* (AC, ARIZ, CS, DS, E, G, GH, ILL, MIN, MO, NMC, NY, P, POM, UC, US). Hidalgo Co., Animas Mts, Lower Indian Creek Canyon, *Wagner 1711* (UNM). Lincoln Co., Ruidoso Creek, 1895, *Wooton s.n.* (US). Otero Co., Tularosa Creek, *Wolf 2820* (BH, CAS, DS, GH, POM). San Migual Co., mouth of Mora River, *Standley 4241* (GH, IND, MIN, MO, NMC, NY, UC). Santa Fe Co., Santa Fe Cañon, *Heller & Heller 3834* (AC, CU, KSC, MIN, NY). Sierra Co., Hillsboro Peak, *Diehl 521* (POM). Socorro Co., Mogollan Mts, *Metcalfe 333* (ARIZ, DS, MIN, NMC, POM, UC, US). Valencia Co., Zuni Mts, *Riffle 646* (UNM).

Oenothera pubescens, a permanent structural heterozygotic species, is one of the few species of *Oenothera* with a natural distribution that extends over both Americas, from Arizona and New Mexico (and possibly southeastern California) across the highlands of Mexico and Central America up to the province of Junín in the Andes of Peru. Unlike the lowland species *O. laciniata,* it occurs strictly in montane sites.

In most recent treatments, *O. pubescens* has been considered a subspecies of the North American *O. laciniata* (Munz 1965; Dietrich 1977), based on the overall morphological similarity between these two species. The crossing experiments made at the Botanical Institute of the University of Düsseldorf (Stubbe & Raven 1979) revealed that *O. pubescens* is a distinct species, however, and does not have an especially close relationship to *O. laciniata.* Hybrids between *O. pubescens* and species of ser. *Raimannia* can be obtained, but they are highly sterile and apparently unable to form viable seed (Stubbe & Raven 1979). In contrast, hybrids in both directions between *O. breedlovei* and *O. pubescens* are a normal green and form normal seeds when open pollinated. The hybrids between *O. breedlovei* and *O. pubescens* examined cytologically in meiotic metaphase I form a ring of 12 chromosomes and one bivalent (DUSS cult. no. 79–205, 206, 208, 82–709 & 710). *Oenothera breedlovei* and *O. pubescens* as pollen parents produce two morphologically intermediate phenotypes that are only slightly different from each other; therefore, *O. pubescens* probably transmits both complexes through the pollen. However, it could not be shown with certainty that both complexes can also be transmitted through the egg cell.

Generally similar to *O. laciniata* of ser. *Raimannia, O. pubescens* is, like that species, extraordinarily variable, as is suggested by the number of synonyms. Two synonyms are based on plants from Arizona and New Mexico (*O. nyctaginiifolia, Anogra amplexicaulis*), three on plants of Hidalgo, Jalisco, and Nuevo León (*Raimannia confusa, R. colimae, Oenothera pennellii*), and two are based on plants from Ecuador (*O. stuebelii, O. diversifolia*), where the type of *O. pubescens* also was collected.

In order to comprehend the amplitude of variation in *O. pubescens,* a number of forms are noted here, beginning with a phenotype that occurs in the Ecuadorian Andes (e.g., *Lugo 509;* 1972, *Müller s.n.*). These plants have an erect to ascending main shoot, slightly incised cauline leaves, scattered pubescence that does not obscure the green color of the leaves, and petals with an average length of 1.5 cm. In Mexico, some plants, as discussed under *O. breedlovei,* have large flowers. Other Mexican plants (e.g., *Sharp 45491,* Veracruz; *Breedlove 36502,* Durango) are densely strigillose, especially in their lower parts, giving the leaves a grayish green appearance. Furthermore, these plants have relatively deeply divided leaves similar to those which are characteristic of *O. breedlovei.* They differ, however, from *O. breedlovei* by their considerably smaller petals. These plants also behave differently in growth patterns in cultivation (e.g., *Breedlove 36502*). While most individuals of *O. pubescens* that are grown from seed sown in February or March usually flower in June, these densely strigillose Mexican plants develop a large rosette and flower late in the summer or not at all. At the end of the growing season, densely leafy lateral branches sprout from the rosette. If these plants survive the winter, they produce a main shoot that flowers heavily in the second year, followed by lateral branches. Since the dense pubescence that is characteristic of this form does not occur in the inflorescence, plants of this sort cannot be distinguished from others after their rosette and lower stem leaves have withered.

Also noteworthy are the rather small-flowered plants with petals less than 1 cm long with a rounded, rather than truncate or retuse, apex. Often these plants seem to flower in the rosette stage, and then develop only short, lateral flowering shoots. The type of *O. pennellii,* from Nuevo León, Mexico, resembles plants of this kind. Other examples of this phenotype are: *Henrickson 11643* (in cultivation in Düsseldorf) and *Purpus 4624,* both from Coahuila, *Moore 3157* from Hidalgo, and *Pennell 17466* and *17467,* both from Zacatecas, Mexico.

Another group of plants that has an erect or ascending growth habit and subentire, narrow cauline leaves is apparently limited to northwestern Mexico and the southern United States (e.g., *White 3441,* Sonora, Mexico; *Hess 2263,* New Mexico; and 1891, *McDougal s.n.,* Arizona).

Our current level of knowledge suggests that there might be morphologically and geographically distinct units within *O. pubescens,* but it has not been possible for us to specify either their characteristics or their ranges completely. Many of the most prominent features by which they can be recognized are expressed fully only in cultivation, as we have mentioned above. In view of this, and especially in view of the prevalent autogamy in this species, which would tend to perpetuate individual combinations of characteristics, we have refrained from considering formal taxonomic recognition for any of these distinctive populations at this time.

Doubtful and Excluded Names

Oenothera diversifolia Sodiro ex H. Léveillé, Monde Pl. 2(97): 40. 1897, nomen nudum.

Oenothera viscosa Rafinesque, Fl. ludov. 96. 1811, based on *Onagra visceuse* Robin, Flore Louisianaise 491. 1807.—This name probably applies to a species of ser. *Raimannia,* perhaps *O. drummondii* or *O. humifusa.* No authentic specimens have been found. Because the description is too brief, this name cannot be assigned with certainty to any known species of *Oenothera.*

Rose proposed the genus *Raimannia* in 1905, but it was not validly published at that time (ICBN Art. 42.1); consequently, none of the new species and new combinations then included by Rose in his new genus were validly published (ICBN Art. 43.1). These invalidly published names are listed below.

Raimannia Rose, Contr. U.S. Natl. Herb. 8: 330. 1905.

Raimannia colimae Rose, Contr. U.S. Natl. Herb. 8: 330. 1905.

Raimannia confusa Rose, Contr. U.S. Natl. Herb. 8: 330. 1905.

Raimannia curtissii Rose, Contr. U.S. Natl. Herb. 8: 330. 1905.

Raimannia drummondii (Hooker) Rose, Contr. U.S. Natl. Herb. 8: 331. 1905.

Raimannia grandis (Britton) Rose, Contr. U.S. Natl. Herb. 8: 331. 1905.

Raimannia heterophylla (Spach) Rose, Contr. U.S. Natl. Herb. 8: 331. 1905.

Raimannia humifusa (Nuttall) Rose, Contr. U.S. Natl. Herb. 8: 331. 1905.

Raimannia laciniata (Hill) Rose, Contr. U.S. Natl. Herb. 8: 331. 1905.

Raimannia littoralis (Schlechtendal) Rose, Contr. U.S. Natl. Herb. 8: 331. 1905.

Raimannia rhombipetala (Nuttall ex Torrey & A. Gray) Rose, Contr. U.S. Natl. Herb. 8: 331. 1905.

ACKNOWLEDGMENTS

This study was supported by a series of grants to the Missouri Botanical Garden (Peter H. Raven, Principal Investigator) from the U.S. National Science Foundation (DEB–7823400, DEB–8119291, BSR–8214879, BSR–8518906). We thank Professor W. Stubbe for helpful discussions and for providing facilities at the Botanisches Institut at the University of Düsseldorf; W. Behn and O. W. Wasmund for useful discussions; L. Mencke for assistance in the laboratory and garden; and all who contributed seeds for the study (see the names that are mentioned under "Specimens from cultivated plants"). We are grateful to the late Carla Lange for translating the first draft of the introductory text from the German. Thanks also to Susan W. Mill for careful preparation of the maps and entering the manuscript on the word processor; to Peter C. Hoch for much assistance, especially logistics between St. Louis and Honolulu, and for overseeing the preparation of the illustrations and specimens cited under *Oenothera laciniata* in St. Louis; and to John Meyers for preparing the excellent illustrations. We appreciate useful comments from A. C. Smith on J. N. Rose's names in *Raimannia.* We thank C. E. Jarvis for having the original plates of *O. laciniata* prepared. We are grateful to the staff of the Missouri Botanical Garden for providing work space and kind assistance. We also thank the staffs of the following herbaria for the loan of material for this study: A, AC, ALU, AMD, APCR, ARIZ, ASC, ASTC, AUA, B, BAB, BAS, BC, BH, BHO, BISH, BM, BP, BREM, C, CAN, CAS, CM, COCO, COLO, CONN, CS, CU, DAO, DS, DUKE, DUSS, E, ENCB, F, FI, FLAS, FR, FSU, G, GA, GH, GOET, HAL, HBG, HUJ, IA, ILL, ILLS, IND, ISC, K, KANU, KE, KR, KRAM, KSC, KY, L, LD, LE, LINN, LISE, LL, LSU, LY, M, MA, MASS, MEL, MEX, MEXU, MICH, MIL, MIN, MISS, MISSA, MNA, MO, MPU, MSC, MT, MU, NA, NCSC, NCU, ND, NDA, ND-G, NEB, NMC, NO, NOSU, NSW, NU, NY, O, OKL, OKLA, OS, OSH, P, PAC, PAM, PENN, PERTH, PH, PHIL, POM, PRC, PRE, R, RSA, S, SD, SIU, SMS, SMU, SP, SRFA, SRGH, TAES, TENN, TEX, TO, TTC, U, UARK, UC, UMO, UNA, UNCC, UNM, UPS, US, USC, USCH, USF, USFS, UWL, VDB, VPI, VT, W, WIS, WVA, YU, Z.

LITERATURE CITED

Asai, Y. 1973. Notes on some species of *Oenothera* in Japan. Jap. J. Bot. 48: 182–187.

Bailey, F. M. 1913. *Comprehensive catalogue of Queensland plants.* Brisbane: Cumming.

Beadle, N. C. W., O. D. Evans, and R. C. Carolin. 1972. *Flora of the Sydney region.* Sydney: Reed.

Blackall, W. E., and B. J. Grieve. 1965 (Reprint, 1974). *How to know Western Australian wildflowers,* III. Univ. West. Austr. Press.

Britton, N. L., and A. Brown. 1913. *An illustrated flora of the northern United States,* ed 2. New York: Charles Scribner's Sons.

Cleland, R. E. 1960. The S-factor situation in a small sample of an *Oenothera (Raimannia) heterophylla* population. Ztschr. Vererb. 91: 303–311.

———. 1968. Cytogenetic studies on *Oenothera* subgenus (*Raimannia*). Jap. J. Genetics 42: 329–334.

———. 1972. *Oenothera* cytogenetics and evolution. London: Academic Press.

Dietrich, W. 1977 [1978]. The South American species of *Oenothera* sect. *Oenothera* (*Raimannia, Renneria;* Onagraceae). Ann. Missouri Bot. Gard. 64: 425–626.

Dietrich, W., P. H. Raven, and W. L. Wagner. 1985. Revision of *Oenothera* sect. *Oenothera* subsect. *Emersonia* (Onagraceae). Syst. Bot. 10: 29–48.

Dietrich, W., and W. L. Wagner. 1987. New taxa of *Oenothera* sect. *Oenothera* (Onagraceae). Ann. Missouri Bot. Gard. 74: 144–150.

Ellstrand, N. C., and D. A. Levin. 1980a. Recombination system and population structure in *Oenothera.* Evolution 34: 923–933.

———. 1980b. Association of alleles with chromosomal complexes in the permanent translocation heterozygote, *Oenothera laciniata.* Heredity 44: 169–176.

———. 1980c. Evolution of *Oenothera laciniata* (Onagraceae), a permanent translocation heterozygote. Syst. Bot. 5: 6–16.

———. 1982. Genotypic diversity in *Oenothera laciniata* (Onagraceae), a permanent translocation heterozygote. Evolution 36: 63–69.

Fischer, F. E. L., and C. A. Meyer. 1835. *Index secundus seminum, quae hortus botanicus imperialis Petropolitanus pro mutua commutatione offert,* 26–54. Leningrad.

Gregory, D. P. 1963. Hawkmoth pollination in the genus *Oenothera.* Aliso 5: 357–384.

———. 1964. Hawkmoth pollination in the genus *Oenothera.* Aliso 5: 385–419.

Guillarmod, A. J. 1971. *Flora of Lesotho* (*Basutoland*). Munich: Cramer.

Hecht, A. 1950. Cytogenetic studies of *Oenothera,* subgenus *Raimannia.* Indiana Univ. Publ. Sci. Ser. 16: 225–304.

Holsinger, K. E., and N. C. Ellstrand. 1984. The evolution and ecology of permanent translocation heterozygotes. Amer. Naturalist 124: 48–71.

Jean, R. 1975. Les *Oenothera* de Belgique dans l'herbier du Jardin Botanique National. Dumortiera 3: 19–25.

Kent, D. H. 1980. BSBI abstracts from literature relating to the vascular plants of the British Isles 10: 45–47.

Lauener, L. A. 1972. Catalogue of the names published by Hector Léveillé, VII. Notes Roy. Bot. Gard. Edinburgh 31: 401–429.

Lewis, H., and M. E. Lewis. 1955. The genus *Clarkia.* Univ. Calif. Publ. Bot. 20: 241–392.

Linder, R. 1965. In Société d'Etude de la Flore d'Alsace. Issler, Loyson, Walter. La Société, Strasbourg.

McKelvey, S. D. 1955. Botanical exploration of the Trans-Mississippi West: 1790–1850. Jamaica Plain: Arnold Arboretum.

Munz, P. A. 1935. Studies in Onagraceae. IX. The subgenus *Raimannia.* Amer. J. Bot. 22: 645–663.

———. 1965. Onagraceae. N. Amer. Fl. II. 5: 1–278.

Pignatti, S. 1982. *Flora d'Italia,* vol. 2. Bologna: Edagricole.

Raven, P. H. 1964. The generic subdivision of Onagraceae, tribe *Onagreae.* Brittonia 16: 276–288.

———. 1969. A revision of the genus *Camissonia* (Onagraceae). Contr. U.S. Natl. Herb. 37: 161–396.

———. 1970. *Oenothera brandegeei* from Baja California, Mexico, and a review of subgenus *Pachylophus* (Onagraceae). Madroño 20: 350–354.

———. 1979. A survey of reproductive biology in Onagraceae. New Zealand J. Bot. 17: 575–593.

Raven, P. H., W. Dietrich, and W. Stubbe. 1979 [1980]. An outline of the systematics of *Oenothera* subsect. *Euoenothera* (Onagraceae). Syst. Bot. 4: 242–252.

Renner, O. 1917. Versuche über die gametische Konstitution der *Oenotheren.* Z. Indukt. Abstammungs-Vererbungsl. 18: 121–294.

Rose, J. N. 1905. Studies of Mexican and Central American plants. No. 4. Contr. U.S. Natl. Herb. 8: 281–339, pl. 63–72.

Ross, J. H. 1972. *The flora of Natal.* Pretoria: Government Printer.

Rostanski, K. 1982. The species of *Oenothera* L. in Britain. Watsonia 14: 1–34.

Rostanski, K., and G. Ellis. 1979. Evening-Primroses (*Oenothera* L.) in Wales. Nat. Wales 16: 238–249.

Rostanski, K., and W. Forstner. 1982. Die Gattung *Oenothera* (Onagraceae) in Österreich. Phyton (Horn) 22: 87–113.

Séringe, N. C. 1828. *Oenothera.* In *Prodromus systematis naturalis regni vegetabilis,* ed. A. P. de Candolle, 3: 45–52. Paris: Treuttel & Würtz.

Small, J. K. 1896. *Oenothera* and its segregates. Bull. Torrey Bot. Club 23: 167–194.

———. 1913. *Flora of the Southeastern United States,* ed. 2.

———. 1933. *Manual of the southeastern flora.* New York.

Spach, E. 1835. Monographia Onagrearum. Nouv. Ann. Mus. Paris 4: 321–407.

Sprague, T. A., and A. M. Riley. 1921. Notes on *Raimannia* and allied genera. Kew. Bull. 1921: 198–201.

Stafleu, F., and R. S. Cowan. 1981. *Taxonomic literature,* vol. 3: Lh–O. Utrecht: Bohn, Scheltema & Holkema; The Hague: W. Junk.

Straley, G. B. 1977 [1978]. Systematics of *Oenothera* sect. *Kneiffia* (Onagraceae). Ann. Missouri Bot. Gard. 64: 381–424.

Stubbe, W. 1980. Über die Bedingungen der Komplexheterozygotie und die beiden Wege der Evolution komplexheterozygotischer Arten bei *Oenothera.* Ber. Deutsch. Bot. Ges. 93: 441–447.

Stubbe, W., and P. H. Raven. 1979. A genetic contribution to the taxonomy of *Oenothera* sect. *Oenothera* (including subsections *Euoenothera, Emersonia, Raimannia* and *Munzia*). Pl. Syst. Evol. 133: 39–59.

Thellung, A., and F. Zimmermann. 1916. Neues aus der Flora der Pfalz. Repert. Spec. Nov. Regni Veg. 14: 375 (215).

Tobe, H., W. L. Wagner, and H.-C. Chin. 1987. A systematic and evolutionary study of *Oenothera* (Onagraceae): seed coat anatomy. Bot. Gaz. (Crawfordsville) 148: 235–257.

Torrey, J., and A. Gray. 1840. *A flora of North America,* 1: 361–711. New York: Wiley & Putnam.

Towner, H. F. 1977. The biosystematics of *Calylophus* (Onagraceae). Ann. Missouri Bot. Gard. 64: 48–120.

Wagner, W. L. 1983. New species and combinations in the genus *Oenothera* (Onagraceae). Ann. Missouri Bot. Gard. 70: 194–196.

———. 1986. New taxa in *Oenothera* (Onagraceae). Ann. Missouri Bot. Gard. 73: 475–480.

Wagner, W. L., R. E. Stockhouse, and W. M. Klein. 1985. The systematics and evolution of the *Oenothera caespitosa* species complex (Onagraceae). Monogr. Syst. Bot. Missouri Bot. Gard. 12: 1–103.

Wasmund, O. W. 1980. Cytogenetische Untersuchungen zur Systematik einiger Sippen der Subsektion *Euoenothera* (Onagraceae). Schriftl. Hausarbeit Univ. Düsseldorf.

———. 1984. Genetische und zytologische Untersuchungen an verschiedenen Sippen von *Oenothera nutans* Atkinson. Thesis, Düsseldorf.

Zohary, M. 1972. *Flora Palestina,* vol. 2. Jerusalem: Israel Acad. Sci. & Hum.

NUMERICAL LIST OF SPECIES

1a. *Oenothera heterophylla* subsp. *heterophylla*
1b. *Oenothera heterophylla* subsp. *orientalis*
2. *Oenothera cordata*
3. *Oenothera rhombipetala*
4. *Oenothera clelandii*
5. *Oenothera curtissii*
6. *Oenothera grandis*
7. *Oenothera falfurriae*
8. *Oenothera mexicana*
9. *Oenothera laciniata*
10a. *Oenothera drummondii* subsp. *drumondii*
10b. *Oenothera drummondii* subsp. *thalassaphila*
11. *Oenothera humifusa*
12. *Oenothera breedlovei*
13. *Oenothera tamrae*
14. *Oenothera pubescens*

INDEX TO NUMBERED COLLECTIONS CITED

The numbers in parentheses refer to the corresponding species in the text and in the Numerical List of Species presented above. × indicates a specimen intermediate between two taxa, when between species they are usually hybrids.

Bell, C. R. 1809 (9); 2260 (9); 2477 (9 × 11); 6164 (9); 6288 (9); 6428 (9); 6591 (9); 6715 (9); 6876 (9); 7095 (9); 7315 (9); 7929 (9); 8392 (9); 11480 (9 × 11); 11738 (9); 12893 (9).
Bell, W. B. 458 (9).
Benbow 81 (6); 82 (3).
Benedict 3321 (9).
Benke 4287 (6).
Berkner 1162 (9 × 11)
Bicknell 6619 (9); 6620 (9).
Biltmore 1383 (6).
Blain 115 (9).
Blake 1700 (9); 11358 (9).
Blanch 72 (6).
Blankinship 17298 (9).
Blocker 159 (9).
Blum 3470 (9).
Blumer 1559 (14); 3458 (14).
Bollenbach 125 (3).
Bonner 204 (9).
Boon 182 (1a).
Boone 452 (9).
Boufford 5168 (9); 19987 (9).
Boufford & Mitsuta 19887 (9).
Boufford et al. 12775 (9); 18579 (11).
Bouly de Lesdain 63 (9).
Bowers & Morton 43543 (9).
Bozeman 2692 (9); 4765 (9); 5162 (9); 6411 (11).
Brackett 224 (9).
Bradley & Stevenson 3315 (11).
Bragg 79 (3).
Brandegee 218 (10b); 220 (12).
Brass 15221 (9).
Braun 1344 (9); 1423 (9); 4498 (9).
Bray 2 (9).
Breedlove 7212 (14); 10731 (14); 11449 (14); 11563 (14); 13444 (10b); 14340 (14); 15773 (14); 15858 (14); 35872 (14); 36482 (14); 36502 (14); 36512 (14); 36519 (14); 37294 (14); 43362 (12); 44465 (14); 59174A (14).
Breedlove & Axelrod 43231 (12); 43340 (12); 43362 (12).
Briggs 3254 (10a).
Bright 3540 (11).
Brinker 420 (11); 1660 (9); 3619 (9).
Britt 254 (9).
Brooks 4352 (9); 5496 (9); 5812 (6); 5829 (9); 5863 (6); 6012 (6); 6307 (9); 6310 (9); 6311 (6); 6313 (6); 6314 (9); 6315 (6); 6316 (6); 6322 (6); 6324 (9); 6325 (6); 6413 (9); 6452 (9); 6517 (6); 6526 (9); 6563 (9); 6564 (6); 6616 (6); 6621 (9); 6646 (9); 6700 (9); 6762 (6); 6779 (6); 8128 (3); 8129 (3); 8130 (3); 8619 (3); 8620 (3); 8638 (3); 9234 (9); 9285 (9); 9317 (9).
Broome 854 (9).
Brown 2 (9); 663 (9); 720 (11); 1840 (9); 2012 (10a); 2090 (9); 3970 (9 × 11); 6254 (6); 8952 (9); 18816 (9).
Brown & Britton 123 (9); 975 (11).
Browne & Browne 6599 (9).
Bruggen 1668 (9); 1684a (9); 2353 (9); 725 (4).
Brumbach 8303 (9).
Brunton & Varga 5399 (4).
Bryan et al. 15 (6).
Buell 1350 (9).
Bull 113 (6).
Bunyon 2647 (8).
Burgess 418 (11).
Buser 2740 (9).
Bush 245 (4); 551 (6); 820 (3); 7550 (9); 9414 (4); 13376 (9); 14871 (9); 15503 (9); 15630 (9).
Butler 8904 (9).
Callaway 54 (9).
Campbell 108-2 (9).
Cantu 61 (6).
Carter 277 (9); 5848 (10b).
Carter III 485 (9).
Celarier 51–92 (1a × 1b).
Chamberlain 9 (3); 417 (9).
Chambers 850 (10b); 892 (12).
Chandler 206 (9 × 11); 272 (9); 1603 (4); 4447 (9).
Chaney 139 (9).
Chase 1174 (4); 1888 (9); 6043 (7); 8728 (9); 10015 (4); 10103 (4); 10393 (9); 10451 (9); 10571 (9); 11334 (4); 13522 (4); 13714 (9).
Chester 2020 (9).
Chitterden 17 (9).
Chung 1524 (10a).
Churchill 1994 (3); 73–709 (4).
Cibulka 76 (9).
Cinq-Mars 73-16 (9).
Clark 278 (9); 360 (3); 4162 (9); 10789 (9); 13239 (9).
Clayton 2330 (6).
Cleland 424 (9); 452 (6); 468 (10a); 479 (11); 503 (11).
Clemens 11701 (9).
Clement 71–76 (9).
Clements 2861 (9).
Clifton 3041 (6); 3103 (9).
Clute 33 (9).
Cobbe & Cobbe 83 (9).
Coleman 46328 (9).
Collins 105 (9); 385 (9).
Conrad 841 (9); 3076 (14); 3077 (14); 3312 (14); 3313 (14).
Constance 3183 (10b).
Conzatti 2430 (14).
Cooke 1277 (1a).
Cooley 5672 (9 × 11).

Cooley & Brass 3567 (9); 3946 (9).
Cooley & Monachino 5672 (9).
Cooley et al. 7545 (9); 12632 (9).
Coon 089 (14)
Cooperrider 1609 (4); 1813 (4); 1956 (4); 2093 (9); 4063 (9).
Corbin 168 (5).
Core 4339 (9).
Correll 1367 (9); 1985 (9); 2145 (9); 2317 (9); 2391 (9); 15854 (9); 20766 (7).
Correll & Correll 9603 (10a); 10073 (9).
Correll & Johnston 16854 (6).
Cory 3277 (9); 11270 (1a); 13362 (6); 15104 (2); 15105 (2); 15805 (3); 18440 (9); 21934 (1a); 32202 (3); 37210 (9); 37211 (6); 43853 (7); 43853 (9); 49957 (1a × 2); 49957 (1a); 50996 (9 × 10a); 51044 (10a); 51653 (8); 51673 (1a, 6); 52645 (9); 52849 (3); 54072 (6); 54075 (9); 54085 (9); 54162 (9); 55282 (9); 56820 (3); 57238 (8).
Craig & Craig 3377 (9).
Crawford 8 (2); 32 (2).
Creager & Beckner 655 (5).
Crider 31 (9).
Crockett 49A (10a).
Cronquist 4274 (9); 4302 (9); 4951 (9).
Crook & Crook 1068 (9).
Crookshanks 4 (9).
Crosby 4842 (9 × 11).
Crutchfield 565 (9).
Cruz 704 (14).
Cuellar 48 (6).
Cuming 1859 (14).
Curle 102 (10a).
Curtiss 905 (9); 5720 (11); 6880 (5).
Cusick 11850 (9); 12991 (9).
D'Arcy 1459 (11); 2919 (11).
Dansereau et al. 40 (9).
Darby 228 (9).
Davidson 456 (9); 2244 (9); 4177 (9).
Davies 498 (9).
Davis 1397 (9); 3730 (10a); 3832 (10a); 4100 (10a); 4364 (9).
Davis & Davis 10542 (9).
Davis & Mahler 4426 (9).
Davis et al. 8164 (9).
Davy 54 (3).
Deam 950 (9); 5005 (11); 11434 (9); 16963 (9); 16996 (9); 17288 (9); 17986 (4); 21000 (4); 31439 (4); 31865 (9); 35536 (9); 35695 (9); 36296 (9); 38738 (9); 38742 (4); 39356 (4); 39551 (4); 48755 (9); 49026 (9); 52325 (9); 59696 (9).
Dean 80 (9).
Degener 28185 (9).
Degener et al. 12593 (9).
Demaree 3421 (9); 5637 (6); 5862 (9); 7702 (6); 8740 (3); 13062 (3); 14857 (9); 14964 (9); 14980 (9); 15153 (9); 15259 (9); 17032 (9); 17143 (9); 17523 (9); 17592 (9); 17732 (9); 19084 (9); 19167 (9); 20796A (9); 20815 (9); 20887 (9); 20964 (9); 21032 (9); 21041 (9); 21266 (9); 21351 (9); 21374 (9); 22664 (1b); 23550 (9); 25998 (9); 26091 (9); 28864 (9); 29005 (9 × 11); 29022 (9); 30448 (9); 30539 (9); 30578 (11); 30788 (9); 32415 (11); 34657 (9); 36465 (9); 36471 (9); 37762 (9); 40820 (9); 40871 (6); 42266 (9); 43779 (14); 45464 (9); 45783 (9); 47279 (9 × 11); 47478 (9); 47963 (9); 48158 (1b); 48167 (9); 49448 (9 × 11, 9); 49533 (9); 49550 (9); 50050 (9); 50306 (11); 50389 (9); 51655 (9); 52017 (9); 52527 (9); 52542 (1b); 53841 (9); 55330 (9); 55414 (9); 55784 (9); 55876 (9); 55987 (1b); 55989 (9); 57760a (9); 69749 (1b); 70094 (1b).
Demaree & Graham 655 (1b).
Deramus D261 (9); D464 (11); D854 (9 × 11).
Dieckmeter 11 (6).
Dieterle 1851 (4).
Diggs 25 (9).
Dihm 3215 (9).
Dinsmore B1074 (10a).
Dixon 523 (9).
Dooley 12 (9).
Dorantes 397 (10a).
Dorantes & Acosta 2679 (10a).
Dowling 1958 (9).
Drake 29 (6).
Dreher 348 (9).
Dreisbach 952 (9); 981 (9).
Dress & Read 7747 (5).
Drew M16 (9).
Drouit 1787 (4).
Drummond 26 (10a); 53 (2); 74 (1a); 75 (8); 107 (9 × 11).
Drushel 3191 (9).
DuMond 1212 (9).
Duffer 373 (9).
Duke 565 (9).
Duncan 6657 (5); 8215 (9); 9725 (9); 20245 (11); 6025B (9).
Duncan & Adams 19791 (9).
Duval 165 (2).
Dziekanowski et al. 1949 (14).
Eames 5952 (9).
Eames & Austin 8302 (9).
Earle 3546 (9).
Easterly 732 (4).
Eaton 1298 (11).
Edwards 129 (3).
Eickwort 53 (4).
Eig et al. 269 (10a).
Eilers 1778 (4); 2227 (4); 2591 (4); 5063 (4).

Ek 82 (9).
Ellis 64 (9); 28862 (9).
Ellstrand 3b (8); 3c (8); 3d (8); 3g (9); 3h (9); 4b (8); 4c (8); 4d (8); 5c (8); 5d (8); 6c (8); 6d (8); 6e (8); 6g (9); 6i (9); 7b (8); 7c (8); 7e (8).
Engleman 1397 (6); 1405 (9).
Eskew 1618 (9); 4619 (6).
Evans et al. 44470 (9 × 11); 45528 (11).
Evers 3841 (9); 10064 (4); 18355 (9); 18451 (4); 22994 (9); 23736 (9); 23894 (9); 25075 (9); 33716 (9); 39629 (4); 46781 (9); 53558 (9); 64523 (9); 69551 (9); 81141 (9); 82900 (9); 87028 (9); 91619 (4); 94684 (9); 96058 (9); 99127 (9); 108497 (9); 110320 (9); 110397 (9); 113862 (4); 130861 (4).
Eyles 8462 (9).
Eyles & Eyles 8369 (9).
Fagerlund & Mitchell 466 (9).
Faifer 7214 (2).
Faircloth 1907 (9); 3354 (9).
Farlin 12480 (5).
Fassett 3460 (4); 5270 (4).
Faust 110 (9).
Fay 786 (4); 2751 (9).
Fell 57–675 (4); 58–524 (4); f.45540 (4).
Ferguson 6788 (9).
Fernald & Long 10749 (9); 4077 (11).
Fernald et al. 14207 (9).
Fillet 637-164 (14).
Fink 452 (4).
Fisher 14 (1a).
Fitzgerald 287 (9).
Fletcher 1271 (14).
Flores 64 (14).
Flyr 1430 (6).
Fogg 8577 (9); 18773 (9).
Font Quer 460 (10a).
Ford 3875 (11).
Fosberg 15399 (9); 15440 (9).
Foster 55 (9).
Fraser 165 (3).
Fredholm 5141 (9 × 11); 5400 (9).
Friesner 13390 (9); 13588 (9); 22133 (9); 22380 (4).
Fritts 51 (9).
Fryxell 1230 (6); 2510 (1a).
Fuller & Fisher 99 (9).
Gale 137 (11).
Galpin 9447 (9).
Gamez 70 (7).
Gander 7627 (9 × 11).
Garcia 18 (7).
Gates 15172 (6).
Genelle 157 (9 × 11); 647 (9 × 11); 1716 (9 × 11)
Genelle & Fleming 1955 (9); 2852 (9).
Gentry 6 (9); 141 (9); 950 (9).
George 2 (9).
Gershoy 513 (9).
Gerstner 4501 (9).
Gibson 44 (3).
Ghiesbreght 677 (14).
Gilbert & Plymale 716 (9).
Gillespie 206 (7); 207 (6); 52 (9).
Gillespie & Dodd 907 (9).
Gillett 5933 (9).
Gilly 324 (9); 659 (4).
Gleason 8526 (9).
Godfread 5626 (9).
Godfrey 53005 (9); 53212 (11); 56459 (9); 58048 (9); 64265 (5); 65686 (9 × 11); 65763 (9); 69338 (9); 69380 (9); 76646 (5).
Godfrey & Lindsey 56817 (11).
Godfrey et al. 7040 (9).
Goessl 1611 (4); 1881 (4).
Goldblatt 1435A (9).
Gómez-Pompa & Nevling 1256 (10a).
Gongora et al. 8815 (6 × 9).
Goodier 361 (9).
Gooding 548 (14).
Goodman 2160 (3); 5793 (9); 6255 (6).
Goodman & Hitchcock 1330 (14).
Goodman & Lawson 8306 (6).
Gordon-Gray 6358 (9).
Gorgoia et al. 8815 (6).
Gough 224 (6).
Gould 7586 (3); 8558 (9).
Gould & Hycka 8015 (9).
Grace 361 (3).
Greear 66166 (9).
Gregory 283 (8); 284 (6); 416 (10a); 421 (6).
Gregory & Eiten 102 (14); 285 (14).
Gressitt 1772 (10a).
Griffin 1 (1a).
Griffen & Demaree 17 (9).
Grimes 261 (3).
Guerra et al. 46 (6).
Gugich 255 (9).
Gunn 10537 (9).
Hale 47619 (9).
Hall 201 (2); 202 (8); 203 (6); 1222 (9).
Hamerstrom 253 (4).
Hance 4375 (10a).
Hancin 1270 (6).
Hanson 340 (7, 9).
Hansen & Nee 7640 (10a).
Hardin 492 (9).
Hardin & Duncan 14940 (9); 16034 (9).
Harding 396 (1a); 490 (3).
Harger 4543 (9).
Harms 587 (9); 1090 (6).
Harper 4522 (1b); 4545 (9).
Harris 32 (14).

Harrison 1861 (14).
Hartley 1131 (4); 1714 (4).
Hartweg 11 (14).
Hastings & Turner 64-354 (10b).
Hatch 2095 (10a).
Hayden 10090 (4); 11400 (9).
Hebb 325 (9).
Hecht 21 (9).
Heggelund 5 (4).
Heller 945 (9); 1512 (10a).
Heller & Heller 3834 (14).
Henderson 62-191 (9); 62-235 (6); 63-147 (6); 63-572 (6); 63-816 (3); 64-271 (5); 65-133 (9); 65-396 (9); 66-278 (9); 66-306 (9); 67-348 (9); 67-778 (9); 69-172 (9).
Henrickson 1748 (14); 11643 (14); 13455b (14).
Herbst & Takeuchi 6396 (9); 6404 (9).
Hermann 8617 (9); 9035 (4); 9087 (4); 9140 (4).
Herrera & McCart 7696 (6 × 9); 7697 (6 × 9)
Hickok & Fuller 23 (9).
Higgins 6110 (3); 7181 (6); 7243 (3); 7251 (9).
Hill 2582 (9); 6732 (8); 7264 (8).
Hinton 11947 (14); 15014 (14).
Hitchcock 671 (5); 687 (3).
Hixson 37 (9); 68 (9); 140 (3).
Hoisington 319 (3).
Holmes 109 (9).
Hom 19168 (10a).
Hood 901 (9); 3378 (11); 4407 (11); 4669 (9).
Hooper 2 (6).
Hopkins 44 (9); 1397 (6); 1421 (9); 3024 (9).
Hopkins & Demaree 104 (3).
Hopkins & Van Valkenburgh 5783 (9).
Hopkins et. al. 274 (9); 606 (9); 790 (3); 1067 (9).
Horne & Horne 172 (9).
Horr E64 (9); E143 (9).
Hoseney 1 (9).
Hott 36 (9).
House 6425 (4); 22820 (4); 27995 (4).
Howell 328 (9); 8821 (14).
Hubbard 2329 (10a); 2694 (10a).
Hudson 60 (9).
Huenefeld 75 (9).
Hulbert 3695 (9).
Humbles 179 (9).
Hunnewell 5929 (6); 11197 (9).
Hunt 3 (6).
Hunter 114 (11).
Hutchins 694 (3); 1992 (6).
Iltis 920 (9); 9467 (4); 16949 (4); 18523 (3); 21480 (9 × 11); 21646 (9); 23182 (9 × 11); 33159 (9 × 11).
Iltis & Neess 8958 (9).
Innes & Warnock 787 (6); 790 (9).
Irving 93 (3).
Isaac 93 (9).
Isely 3156 (9); 6386 (9).
Iwanicki & Wilkinson 7505-22 (9).
Jackson 177 (3).
Jacobsz 2078 (9).
Jallu 5331 (9).
James 6247 (9); 6675 (9); 6823 (9); N-169 (9).
Jansen & Reichgelt 4056 (9).
Jeffs 55 (6).
Johns 588 (14).
Johnson 27 (9); 90 (6); 1110 (6).
Johnson & Webster 573 (6).
Johnston 5439 (7); 54141 (9 × 10a).
Johnston et al. 3392 (8).
Jones 3A (9); 53 (2); 193 (9); 238 (9); 481 (2); 2705 (9); 4042 (9); 4433 (9); 4469 (9); 4918 (9); 5592 (9); 5936 (9); 11865 (9 × 11); 12020 (9); 15995 (9); 16269 (9); 16650 (4); 16999 (9); 17278 (9); 17300 (9); 18393 (9); 18625 (9); 22258 (9); 27044 (12); 27206 (12); 29534 (7); 29536 (7); 40835 (4).
Jones & Jones 4218 (9); 4243 (9); 4712 (9); 11161 (9); 11284 (9); 22532 (9); 22536 (5).
Jones & Reynolds 11142 (9); 11723 (9).
Jones et al. 16879 (9).
Jones, M. E. 232 (14); 24098 (10b).
Jotter 3878 (4).
Justice 192 (9).
Kanai et al. 10092 (9).
Kearney & Peebles 9819 (14).
Keeler 1870-1876 (5).
Keever 309 (9).
Keil 7465 (9).
Kellogg 25889 (9).
Kelting 200 (3); 238 (6); 249 (3).
Kenoyer 422 (14).
Kern & Reichgelt 5062 (9); 5202 (9).
Kew 209-10 (9); 209-12 (9); 210-02 (9).
Kiener 14702 (9); 21497 (3).
Killick 2307 (9).
King 88 (9); 2026 (14); 6399 (9).
Klatmough 492 (10a).
Knight 86 (1a).
Koch 1807 (3); 3791 (9).
Koehler 48 (9).
Koelling 881 (9).
Koelz 119 (14); 13192 (3).
Kolle 1413 (3).
Kral 2050 (9 × 11); 4191 (9); 8428 (9); 12109 (1b); 25190 (14); 25247 (14); 28167 (9); 30310 (9); 45316 (1b); 45322 (9); 45642 (9); 45955 (9); 46185 (9); 46243 (1b); 46366 (1b); 46777 (1b); 50093 (1b); 52542 (9); 55663 (9 × 11); 55663 (9).
Kral & Demaree 30720 (9).
Kral & Knott 4243 (9).
Kramer 4602 (10a).
Kurz 1601 (5).

Laessler 171 (5).
Laing 988 (9).
Lakela 24957 (11); 25145 (9); 29847 (11).
Lammers 400 (4).
Larsen 929 (11); 1081 (11).
Lassetter 210 (9).
Latham 15 (6).
Lathrop 405 (9); 2019 (9).
Laughlin 1344 (9).
Launchbaugh 31 (9).
Lawson & Massey 293 (6).
Lazor 3194 (9).
LeSueur 319 (6); 320 (10a × 6); 321 (6); 804 (14); 807 (14); 809 (14).
Leavenworth 244 (14).
Lehio 15641 (9).
Leinkram 20252 (10a).
Lemaire 2319 (9).
Leon 11649 (11).
Leonard & Killip 540 (9).
Leonard & Radford 2169 (10a).
Leonard et al. 1761 (11).
Lester 70 (9).
Lewis 1509 (9).
Liebmann 3251 (14).
Lieu 2544 (9).
Lindayen 193 (4).
Lindheimer 35 (1a); 53 (10a); 56 (1a); 69 (10a); 810 (9).
Lipscomb 726 (3).
Little 104 (6); 164 (9).
Loens 1345 (9).
Lonard 8783 (9).
Long 97 (3); 479 (4); 23582 (4); 26641 (4); 31396 (9); 39994 (9); 48786 (4); 49085 (4); 49974 (9); 59458 (9).
Long & Lakela 27578 (11).
de Los Santos 39 (6).
Lot 1299 (10a).
Lousley 2180 (9).
Lucas 14203 (7).
Lucas et al. 14203 (9).
Lundell 10805 (8 × 9); 13054 (3); 13905 (3).
Lundell & Lundell 8356 (9); 8838 (8); 10849 (6); 10855 (6); 10924 (9).
Luteyn 2060 (4).
Mackenzie 30 (7); 124 (6); 288 (9); 4455 (11).
Mahler 5163 (10a); 5244 (7); 5348 (8 × 9); 6488 (9).
Makino 42840 (9); 42843 (9); 42844 (9).
Mann 66 (9).
Manning 55104a (14).
Manuel 168 (11).
Marsh 2031 (9).
Martin 299 (9).
Mason 1352 (4).
Massey 500 (9).
Massey & Massey 3698 (9).
Mattison 2959 (9).
Maupin 578 (9).
Maxon 7 (9).
Maysilles 7326 (14).
McBryde 3034 (6).
McCart 1102 (3); 6314 (7); 6378 (9); 6416 (9); 6455 (9); 11215 (11).
McCoy 796 (3); 2624 (9); 6003 (9).
McCrary 81 (3).
McCree 799 (9).
McCullough 59 (6).
McDaniel 4251 (9).
McDonald 271A (9); 5104 (10a).
McDougall 178 (4).
McGrary & McGrary 37 (6).
McGregor 21 (9); 4817 (9); 12482 (6); 12489 (3); 12648 (6); 15850 (3); 16876 (9); 16914 (6); 17213 (9); 19331 (9); 24802 (9); 24811 (9); 24861 (9); 24874 (9); 25467 (6); 25512 (6); 25535 (6); 25624 (3); 25634 (3); 26327 (3); 27260 (6); 28976 (9); 29012 (9).
McGregor & Bare 1678 (3); 1724 (3).
McInteer 1753 (9).
McKellar 28 (5).
McNiel 33 (9).
McVaugh 5738 (11); 12379 (1a × 2); 12379 (1a); 16882 (14).
Means, Jr. 3339 (9); 3395 (9).
Measeles 17 (9).
Medrano 521 (10a).
Meebold 3444 (10a); 27700 (5).
Mekee 9495 (10a).
Merrill 272 (9); 739 (4).
Mertins 1008 (9).
Merxmüller 26545 (9).
Metcalfe 196 (14); 333 (14); 1054 (14).
Metz 578 (8).
Meyer & Mazzeo 1089 (9).
Milby 18 (9).
Miller et al. 20725 (9).
Millward 74 (10a).
Mohr 524 (1b).
Moldenke 1038 (11); 1074 (9); 26637 (9 × 11).
Moldenke & Moldenke 26418 (9).
Molina 15491 (14).
Montz 496 (9 × 11); 822 (9); 1035 (9).
Moody 110 (9).
Moore 12 (9); 713 (9); 2076 (9); 2954 (9); 25669 (4); 56-114 (1b); 56-134 (1b).
Moran 7016 (10b).
Morfino (10a).
Morley 72 (9).
Morrill 446 (9).
Mosquin 5566 (8).
Mosquin & Mosquin 5464 (9); 5808 (9).
Mostert 632 (9).

Moyer 229 (9).
Mühlenbach 63 (6); 942 (4); 1018 (1a); 3316 (9).
Muenscher 17257 (9).
Muenscher & Winne 16639 (9).
Mulcahy 131 (9).
Muller 8518 (3).
Munz 1946 (9); 6127 (9); 6127 (9); 13333 (2); 13335 (2); 13344 (1a); 17531 (3).
Munz & Gregory 23432 (6); 23450 (10a); 23511 (3).
Murray 45528 (9 × 11).
Murtishaw 185 (9).
Nash 225 (9); 515 (9 × 11).
Nelson 39 (6); 379 (3); 6112 (14).
Nighswonger 869 (3).
Nixon G12 (9); 4421 (9).
Nixon & Sullivan 461 (9).
Nixon et al. 7107 (9).
Noble et al. 11561 (9).
Norton 158 (9).
Notzon 27 (6 × 9).
Olney 8 (9 × 10a).
Omar 32 (9).
Oosting 95 (9).
Orcutt 5536 (10a × 6).
Osborn 1163R (9).
Over 16134 (3).
Ownbey 743 (9).
Palmer, E. 40 (14); 132 (6); 155 (14); 207 (14); 255 (10a × 6); 282 (10a); 343 (10a).
Palmer, E. J. 301 (9); 550 (6); 5007 (10a); 5016 (9); 5097 (9, 9 × 10a); 5991 (4); 7302 (9); 7647 (6); 7648 (9); 8325 (3); 9248 (7, 10a); 9626 (2); 10054 (6); 10334 (3); 11290 (7); 11698 (8); 11965 (10a); 33731 (9); 40566 (4); 41956 (3); 43687 (4); 55695 (9); 62155 (9); 65744 (9).
Palmer, E. J., & Steyermark 40911 (9).
Pammel & Reppert 1219 (4).
Parish 6503 (9).
Parkey 77 (3).
Parks 124 (9).
Parks & Cory 1502 (2); 15101 (2); 15103 (2).
Parks et al. 271 (7).
Parmelee 764 (3).
Parry & Palmer, E. 253 (14).
Patman 1067 (9 × 11).
Payton 86 (9).
Pearce 685 (9); 784 (3); 2640 (6).
Peebles et al. 8873 (14).
Pena 8 (7).
Pennell 99 (9); 17139 (14); 17466 (14); 20122 (14).
Pepoon & Barrett 5424 (9).
Perdue 664 (9).
Perino & Perino 416 (9).
Perry & Myers 968 (9).
Pfeifer 2375 (9).
Pierce 314 (4).
Pineda 787 (14); 877 (14).
Pinkava et al. 13762 (14).
Pipley 14668 (14).
Pittillo 117 (9).
Pohl 4906 (6, 9); 4965 (3); 9360 (9).
Pollard 810 (9); 1020 (9 × 11).
Pont 669 (9).
Popowsky 25 (9).
Porter 307 (10b); 1846 (9).
Powell 16012 (3).
Pratt 76 (9).
Pringle 10314 (14).
Proder 43 (3).
Pryor 273 (3).
Pullen 6839 (9); 6868 (9); 7082 (9); 64476 (9); 64737 (9); 67104 (9); 69369 (9); 70115 (9); 70152 (9); 70285 (9); 70460 (9).
Purpus 4624 (14).
Quarterman 5260 (9).
Quibell 5804 (9).
Radford 9193 (9); 10571 (9); 10683 (9); 10869 (9); 11329 (9); 12697 (9); 12769 (9); 13503 (9); 13620 (9); 14907 (9); 20342 (9); 20645 (9); 20756 (9); 20906 (9); 21299 (9); 22103 (9); 31964 (9); 32263 (9); 32335 (9); 33015 (9); 33651 (9); 33760 (9); 33918 (9); 36975 (9); 44784 (9).
Radford & Stewart 339 (9).
Raffaelli 35 (6).
Ramirez et al. 8783 (6).
Ramsey et al. 277 (5); 6794 (9).
Randel 49 (9).
Randolph & Merriman 282 (9).
Raper 27 (9).
Rapp 395 (9).
Ratledge & DeSelm 30754 (9).
Raven 20457 (10a); 26528 (3).
Raven & Gregory 19215 (6); 19395 (9); 19440 (9); 19474 (3); 19489 (3); 19490 (9).
Reddin 25 (9).
Redfearn 3591 (9); 5732 (9).
Redfearn & Houk 8166 (9).
Redfield 9 (14).
Reed 1258 (3); 3286 (3); 3329 (9); 3384 (6); 100083 (9).
Reeves 178 (1a); 2009 (6).
Reidel 44-12 (9).
Rendle 682 (10a).
Repton 5306 (9).
Reverchon 245 (1a); 295 (1a); 297 (6); 905a (6); 2002 (3); 3561 (1a); 3757A (7).
Rhodes 12323 (9); 12409 (9); 14629 (9).
Rice 39 (9).
Richards 2986 (9); 3246 (9).
Rickey 526 (9).

Ridgway 2699 (9).
Riffle 646 (14).
Rill 1659 (4); 3083 (9); 3345 (9); 4095 (4).
Robbins 2398 (9); 3087 (3).
Robinson 194 (9).
Rochat & Sinclair 50 (9).
Rogers 6351-D (9 × 11); 6668-A (9); 9199 (1b).
Rodgers & Mullens 67030 (9).
Rohrer 28 (9).
Rood 649 (9).
Roos 3885 (9).
Rose 2133 (13).
Rose & Hay 5636 (14).
Rose & Painter 6451 (14).
Rose-Innes 41005 (6).
Ross 37 (9).
Rossbach & Murphy 2384 (9).
Rosson 1740 (6).
Rowell, Jr. 4260 (3); 5804 (6); 8003 (9); 10082 (9); 16011 (6); 5376A (9).
Runyon 539 (7); 1844 (7); 2486 (7); 2569 (6); 2646 (7); 2653 (9); 3111 (7); 4922 (7); 4293 (7); 6042 (7).
Rush 150 (9); 444 (9); 493 (3); 1556 (10a).
Rutland 145 (9); 205 (9); 2059 (9); 2300 (9); 2484 (9); 2774 (9).
Rutter et al. 471 (10a × 6).
Rydberg 109 (9); 703 (9); 1510 (3).
Rydberg & Imler 125 (9); 1323 (3).
Rzedowski 20008 (10a); 20935 (14); 21907 (14); 31445 (14).
Saenz 65 (9).
Sampson 244 (10a).
Sancedo 152 (14); 169 (14).
Sanders 140 (9); 156 (1a).
Santarius 2190 (14); 2198 (14); 2235 (14); 2238 (14); 2326 (14).
Sargent 6169 (11).
Sauer 3411 (10a).
Schaeffer 45374 (9).
Schallert 3686 (9).
Schiede 529 (14); 532 (10a).
Schmidt 1806 (9).
Schmitz 39 (9).
Schulz 308 (9).
Schuster A-7608 (9).
Seaton 177 (14).
Seigler 5051 (9).
Semple 539 (9).
Serna 23 (6 × 9).
Setser 494 (9); 675 (9).
Seymour 16776 (9).
Shanks et al. 4199 (9); 14303 (9).
Sharp 45379 (14); 45577 (14).
Sharp & Hesler 1012 (9).
Sharp & Jones 28221 (9).
Sharp et al. 3738 (9); 12133 (9); 12166 (9); 12196 (9); 12496 (9); 12614 (9); 12881 (9); 13007 (9).
Sharsmith 5402 (9).
Shaw & Allred 2007 (9); 2016 p.p. (7); 2016 p.p. (9); 2018 p.p. (8); 2018 p.p. (9); 2020 p.p. (7); 2020 p.p. (9); 2021 p.p. (7); 2021 p.p. (8); 2021 p.p. (9).
Shehane 32 (9).
Shildneck C-6902 (9).
Shimizu 13947 (9).
Shinners 2606 (6); 10201 (3); 10872 (9); 11123 (9); 13983 (9); 14637 (9); 17655 (6 × 9); 17712 (7); 17812 (7); 18595 (6); 20281 (9); 22587 (9); 22802 (9); 22847 (9); 24073 (1a); 25976 (9); 26300 (6); 26407 (6); 26431 (3); 26747 (9); 26941 (9 × 11); 27187 (9); 28059 (9 × 11); 28279 (9); 28543 (9); 29214 (9); 29610 (9 × 11); 29752 (9); 30975 (9); 31210 (9); 31809 (6); 31843 (1a); 33040 (6).
Shreve 1640 (6).
Silberhorn 3771 (9).
Simpson 5056 (10a).
Sinclair 0-3 (9).
Sinclair et al. 1527 (9).
Skehan 83 (9).
Skroch PE112 (4).
Small 8526 (10a).
Small & DeWinkeler 9465 (10a).
Small & Small 6941 (11).
Small & Wherry 11918 (9).
Smith 17 (6); A-64 (9); 165 (9); 268 (9); 404 (9); 412 (9); 523 (9); 594 (3); 1026 (3); 1569 (5); 3465 (9); 7462 (4); 8275 (4).
Smith & Hodgdon 4085 (9).
Sorenson 1681 (4).
Soto & Horvitz 9 (10a).
Sparling 838 (9).
Sperry T603 (9).
Standley 4241 (14); 59262 (14); 61493 (14); 65762 (14).
Stanford 2234 (9).
Steibel 8532 (9).
Stellar 9 (6).
Stephens 3396 (9); 4214 (9); 4455 (9); 4525 (9); 10461 (9); 11123 (9); 11212 (6); 11281 (6); 15473 (3); 15516 (3); 15768 (3); 16183 (3); 17166 (3); 18997 (3); 19154 (3); 20401 (6); 20494 (9); 20694 (6); 20779 (9); 20900 (9); 21591 (9); 24418 (3); 28175 (3); 30554 (9); 30756 (9); 34587 (3); 47389 (9); 47638 (6); 47639 (9); 47785 (6); 49242 (3); 53822 (9); 53965 (6); 56429 (9); 57699 (9); 60548 (3); 62118 (3); 64433 (9); 64887 (9); 65232 (6); 65233 (9); 65294 (9); 65395 (9); 72996 (6); 74738 (6); 74739 (9); 74810 (9); 74828 (9); 74873 (9); 76475 (6); 76552 (6); 79036 (9); 79083 (9); 80326 (6); 80675 (3); 80966 (3);

81021 (3); 81703 (3); 84487 (9); 84903 (9); 87695 (3).
Stephens & Brooks 24058 (6); 24183 (3); 24265 (3); 24351 (3); 24914 (3); 24998 (3); 25033 (3); 32143 (9); 34216 (3); 34293 (3); 38613 (9).
Stern 274 (3).
Stevens 95 (9); 618 (9); 1035 (3); 1194 (6); 1195 (6); 1392 (9); 3021 (6).
Steward 3084 (10a).
Stewart 47 (6); 160 (9).
Steyermark 35948 (14); 40803 (4).
Stone 262 (9 × 11); 497 (9).
Straley 888 (9).
Straw 1808 (14).
Straw & Forman 1469 (14); 1918 (14).
Strey 6963 (9); 7783 (9).
Strother 90 (6).
Stübel 161b (14).
Stuessy 1057 (14).
Swink 64 (9); 602 (4); 2319 (4).
Tatnall 3675 (9).
Taylor 49-1228 (9).
Taylor & Baalman 2618 (6).
Taylor & Taylor 13255 (10a).
Temple 2935 (9); 4782 (9); 5047 (9); 8109 (9); 8863 (9).
Tharp 860 (1a); 2917 (1a × 1b); 44140 (9); 49138 (8 × 9, 9 × 10a).
Tharp & Gimbredge 52631 (1a).
Theron 1146 (10a).
Thieret 8839 (9 × 10a); 8935 (9); 10286 (9); 22433 (9 × 11); 29168 (11); 29169 (9); 30305 (1a); 31220 (1a); 31638 (1a); 32952 (9).
Thomas 1002 (9); 9942 (9).
Thomas et al. 7021 (9).
Thompson 302 (9); 558 (9).
Thorne 5744 (5); 13655 (4); 20166a (10a).
Thorne & Davidson 16410 (5).
Thorne et al. 14115 (9).
Thorrow 9054 (4).
Timmons 345 (9).
Todd 281 (11).
Tolstead 7025 (6); 7485 (3); 7507 (9).
Tombe 515 (9).
Toole 11 (1a).
Tosh 887 (9).
Tracy 131 (9); 6614 (5); 7601 (11); 7820 (3); 8719 (11); 9222 (10a); 9274 (9).
Traverse 388 (7); 968 (10a); 1167 (10a); 1188 (10a); 1396 (9).
Trécul 1438D (1a).
Trew 199 (9).
Trioani & Prina 8226 (9).
Troiani & Steibel 7733 (9).
Trott 134 (9 × 11).
True 128 (9).
Tsoong 4337 (10a).
Tucker 3838 (9); 5731 (1b); 7207 (3); 8079 (1b).
Tunnell 15 (9).
Turley 74 (6 × 9).
Turner 1793 (9); 4964 (2).
Tyson 63 (5); 75 (5); 7093 (9).
Ugent & Flores 2185 (14); 2343 (14); 5798 (14).
Underwood 267 (4).
Ungar 769 (9).
Uttal 8696 (9); 9424 (4); 11019 (9).
Van Vleet 51 (9).
Vandergriff 31 (3).
Ventura 4689 (14); 12553 (10a).
Villarreal 46 (7).
Voss 2856 (4); 7569 (4).
Wadmond 3281 (4).
Wagenknecht 2456 (6); 2471 (9); 3553 (9).
Wagner 358 (14); 1711 (14).
Wagner & Solomon 4276 (14).
Wagner et al. 5544 (9).
Walker 16 (6); 17 (1a); 92 (9); 3420 (9).
Waller 154 (1a).
Wallis 2916 (9); 3799 (6 × 9); 6643 (9); 6694 (9); 6961-1 (9); 7449 (9); 7450 (3); 8129 (9); 8447 (9).
Ward A-34 (9 × 11); 3763 (9); 7836 (10a); 7837 (10a × 6).
Warnock 14633 (6); 15425 (3); 20548 (9); 47269 (9).
Waterfall 427 (6); 1474 (3); 2056 (9); 2958 (9); 16934 (6); 17461 (6).
Waterfall & Coryell 9918 (6).
Waterfall & Wallis 13462 (14).
Watson 60 (6).
Watt 733 (9 × 11).
Watters 1123 (9).
Weatherby & Griscan 16594 (9).
Weatherwax 3667 (4).
Weber 32 (6); 338 (14); 392 (9); 14448 (6).
Webster 2771 (9).
Webster & Rowell, Jr. 7091 (2).
Webster & Webster 7076 (9).
Wehmeyer 79 (4).
Welch 1844 (4).
Wells 2849 (9).
West & Arnold 171 (9).
West 1129 (9).
Weston 2106 (9); 2306C (9); 5686 (9).
White 3441 (14); 6745 (10a).
Whitehouse 9646 (6); 9782a (6); 9816 (9); 10009 (9); 10062 (3); 10478 (3); 15009 (9); 16778 (3); 18137 (8 × 9); 18138 (8); 18255 (6); 19828 (8); 19894 (9); 20876 (9 × 10a); 21369 (1a).
Wiedeman 367 (9).
Wiegand & Manning 2208 (5).

INDEX TO SCIENTIFIC NAMES

Accepted names are in roman type; the main entry for each is in **boldface.** Synonyms are in *italics.*